Renewable Energy – The Facts

Renewable Energy – The Facts

Dieter Seifried and Walter Witzel

publishing for a sustainable future

London • Washington, DC

First published in 2010 by Earthscan

Original German version published as: Walter Witzel and Dieter Seifried (2007) *Das Solarbuch. Fakten, Argumente und Strategien für den Klimaschutz*, 3rd edition, Ökobuch Verlag.

Published by Energieagentur Regio Freiburg, Freiburg/Germany
www.energieagentur-freiburg.de
info@energieagentur-freiburg.de

**Energieagentur
Regio Freiburg**

Earthscan Ltd, Dunstan House, 14a St Cross Street, London EC1N 8XA, UK
Earthscan LLC, 1616 P Street, NW, Washington, DC 20036, USA
Earthscan publishes in association with the International Institute for Environment and Development

For more information on Earthscan publications, see www.earthscan.co.uk or write to
earthinfo@earthscan.co.uk

ISBN: 978-1-84971-159-3 hardback
ISBN: 978-1-84971-160-9 paperback

Typeset by FiSH Books, Enfield
Cover design by Yvonne Booth
Translated by Petite Planète Translations

A catalogue record for this book is available from the British Library

Library of Congress Cataloging-in-Publication Data

Seifried, Dieter.
  Renewable energy : the facts / Dieter Seifried and Walter Witzel.
    p. cm.
  Includes bibliographical references and index.
  ISBN 978-1-84971-159-3 (hardback) – ISBN 978-1-84971-160-9 (pbk.) 1. Renewable energy sources.
  I. Witzel, Walter. II Title.

TJ808.S54 2010
333.79′4–dc22   2010020265

At Earthscan we strive to minimize our environmental impacts
and carbon footprint through reducing waste,
recycling and offsetting our CO emissions, including those
created through publication of this book. For more
details of our environmental policy, see www.earthscan.co.uk.

FSC
www.fsc.org
MIX
Paper from
responsible sources
FSC® C013604

Printed and bound in the UK by CPI Antony Rowe.
The paper used is FSC certified.

Contents

List of Figures

Foreword

There are dark clouds on the horizon. Climate change – long researched, discussed and denied – is increasingly making its presence felt. Drawn up by more than 2000 climate researchers from around the globe, the International Panel on Climate Change's (IPCC) 2007 report has a clear message: the Earth will inevitably heat up by more than 2°C above the temperature of the preindustrial age. Additional warming would have enormous consequences for mankind and the environment, and a global economic crisis can only be avoided if the global community works closely together.

'The time for half measures is over', former French President Jacques Chirac once said, commenting on the challenges of climate protection. 'It is time for a revolution – an awareness revolution, an economic revolution, and a revolution of political action.'

Unlike the three industrial revolutions (the first with the steam engine, loom and railways; the second with crude oil, cars and chemistry; and the third with information technology and biotechnology), the fourth industrial revolution will have to be part and parcel of a transition to a solar economy – and it will have to be a global revolution.

Despite all the talk, global energy consumption continues to rise from one year to the next. Industrial nations have only adopted modest climate protection policies, and energy consumption is skyrocketing in the most populous developing nations of China and India. We are called on to cut global greenhouse gas emissions in half by 2050; at the same time, poor countries continue to

fight for their right to economic development. Therefore, our global switch to a renewable energy supply must be based on a dual strategy: greater energy efficiency and the fast development of renewable energy.

The dark clouds on the horizon do indeed have a silver lining of sorts. Behind them is a blue sky and a shining sun. The fourth industrial revolution of efficiency and solar power will make our energy supply safer. No longer will we fight for oil, and the battle against poverty will be won. Millions of new jobs will be created, and national economies and consumers will face less of a financial burden. The only thing to fear is inaction.

But the fear of inaction should be taken seriously. The main energy efficiency technologies and eco-efficient products – from cars that get 80 miles per gallon to cogeneration systems and homes that produce more energy than they consume – are already available. Seifried and Witzel show a wide range of these convincing options in practice and discuss the political reasons for society's reluctance to become more efficient.

In *Renewable Energy – The Facts*, the authors concentrate on the second major challenge we face: covering all of our (drastically reduced) global energy consumption with renewables. They convincingly show the great technical and economic potential of solar energy alongside that of wind, water and biomass, each of which can be considered indirect solar energy.

And that's not all. They also show that a narrow focus on technical potential is near-sighted. The drastic structural change in our energy sector and society will only come about if society undergoes an innovation process. In addition to technologies, this process requires the will to march on into sunnier days. It also requires proper institutional and market conditions – and different consumer behaviour, both in terms of purchases and product use.

The questions seem to be endless, but the answers are provided in the book you hold in your hands. *Renewable Energy – The Facts* is a manual for the fourth industrial revolution.

Rainer Griesshammer
Rainer Griesshammer is a member of the board at the Institute of Applied Ecology and a member of the German Advisory Council on Global Change.

Preface

'Renewables are the way of the future' – 20 years ago, this was a minority opinion. Back then, our energy supply came from fossil sources (coal, oil and gas) and from nuclear power. Power providers did not believe that solar energy could ever make up a large share of the pie and merely spoke of it as the 'spare tyre', which was good to have on board, but not something you would want to rely on all the time.

Over the past few years, opinions have begun to change. Markets for renewable energy sources are booming around the world. At the same time, the negative effects of our fossil-nuclear energy supply become clearer all the time:

- The dramatic impact on the climate of our uninhibited consumption of fossil energy is causing glaciers and polar ice to melt at rates previously unimagined. Ironically, the deserts are also expanding. Higher temperatures foster the spread of malaria and cholera, and extreme weather events, such as the European heatwave in the summer of 2003 and Hurricane Katrina in 2005, are becoming common. The warnings from researchers about the catastrophic consequences and the tremendous costs of climate change are only becoming more urgent. For instance, in a study published in October 2006, Nicholas Stern, the former chief economist at the World Bank, argued that climate protection is the best economic policy. While a lack of effective climate policies could cause damage amounting to up to 20 per cent of global gross domestic product (GDP), Stern calculated that proper climate protection would only cost 1 per cent of global GDP.[1]
- Crude oil and natural gas are becoming scarcer. Prices skyrocketed in 2008 leading up to the economic crisis, while the war in Iraq was a reminder that most of the world's oil reserves are in an unstable part of the world.
- The reactor disaster in Chernobyl (1986) tragically demonstrated that there is no such thing as safe nuclear power. Indeed, mishaps continue to this day, such as in the summer of 2006 in Forsmark, Sweden, and Biblis, Germany. Furthermore, we still do not know how to safely dispose of nuclear waste, which is why we need to stop making it as soon as possible.

These and other reasons clearly illustrate that our fossil/nuclear energy supply is not sustainable and has no future. At the same time, we are currently witnessing the beginning of the Solar Age and a boom in renewables, though perhaps 'witnessing' is not the right word – we are bringing this change about ourselves. Obviously, solar power is not a marginal player. Instead, it is the only sustainable energy source we have and will be a central pillar of our future energy economy alongside prudent energy consumption.

The trends over the past few years leave room for no other conclusion; solar energy is no longer a marginal player.[2] In 2006, the number of solar arrays installed in Germany crossed the threshold of 1 million. In only seven years, from 1999 to 2005, the industry increased its sales more than tenfold, equivalent to average annual growth of around 50 per cent. In 2005, 45,000 people were employed in the solar sector, which posted €3.7 billion in revenue. By 2020, that figure is expected to increase another sevenfold.

Wind power has grown even faster. Policies in the 1990s got things going, bringing about increasingly powerful wind turbines. For many years, Germany was the world's leader in wind power and was only overtaken by the US in 2008. At the end of 2008, Germany had installed a total capacity of 23,903 megawatts (MW) of wind power. The 20,301 wind turbines in the country generated 40.4 terawatt-hours (TWh) of wind power that year, equivalent to 7.5 per cent of Germany's power consumption. The figure from 2006, only two years earlier, was 5.7 per cent; that year, wind power overtook hydropower as the biggest source of renewable energy.

Nowadays, the payback from policies to promote wind power is clear. German firms are global market leaders. Modern wind turbines are being exported in large numbers because in good locations wind power is cheaper than power from conventional central plants. At the end of 2007, some 90,000 people were employed in the German wind power sector.

Long overlooked, biomass recently moved to centre stage. A number of communities heat new buildings with renewable wood, and wood pellets ovens for detached homes and multi-family units have become genuine competitors for oil and gas heaters. Within just three years, the number of these environmentally friendly boilers rose tenfold. In addition, a growing number of farmers are now growing energy crops. Plantations of rapeseed are a source of additional income alongside biogas digesters.

All of these steps go in the right direction in our opinion, and they are all the results of government policy, such as Germany's Renewable Energy Act (EEG). But Germany is not a special case. A number of countries have adopted similar policies, called feed-in tariffs (FITs). Some 60 countries worldwide have adopted FITs, making it the leading policy instrument to promote renewables worldwide.

Wind power continues to boom worldwide (see www.ewea.org/fileadmin/ewea_documents/documents/press_releases/2009/GWEC_Press_Release_-_tables_and_statistics_2008.pdf). For instance, in 2008, installed wind power capacity rose by some 30 per cent, while the grid-connected photovoltaics (PV) capacity grew by more than 70 per cent.[3] Overall, a total investment of €120 billion (2008) underscores the growing economic importance of the sector.

Crucially, China, the most populous country in the world, has set some ambitious targets for itself. By 2020, renewables are to make up 15 per cent of the country's power consumption. In particular, China installed some 13 gigawatts (GW) of wind capacity in 2009 alone, bringing it more than halfway to its target of 20GW by 2020 – and making China the global wind leader for that year.[4] China also has ambitious plans for other renewable sources of energy, which all goes to show that renewables are a genuine option for developing and newly industrialized countries.

Though the US did not ratify the Kyoto Protocol, more than 300 mayors – from Chicago to New York, Los Angeles and New Orleans – have stated their support for the treaty.[5] And though former President George W. Bush came from the oil sector and was surrounded by consultants from the oil industry, renewables boomed during his administration more than ever before as the country worked to make itself less dependent on foreign energy imports.

Clearly, energy policy is in a transitional period. Renewables are quickly becoming more important. In this book, we navigate our readers through this process and provide them with facts and good reasons for this change. We also present strategies for the quick transition to the Solar Age:

- The book first provides information about the many ways that solar energy can be used. We start with the direct use of solar energy: solar thermal and PV. The former creates heat; the latter, electricity (Chapters 2–4). The sun is also the engine behind our climate; wind, clouds and rain are the result of insolation. Likewise, plants (biomass) could not exist without light. Biomass, wind power and hydropower are therefore thought of as indirect ways of using solar energy. Finally, geothermal is yet another renewable source of energy (Chapters 5–7). We round off this presentation of energy sources with an overview of new energy technologies often mentioned in the context of renewable energy, such as fuel cells (Chapter 8).
- The second part of the book focuses on the overall potential of solar energy. We discuss not only the possibilities of various types of solar energy, but also how they are currently used in Germany, Europe and worldwide. A scenario for the expansion of renewables illustrates our future prospects (Chapter 9). A number of arguments against the expansion of renewables are also repeatedly voiced in the debate about our future energy supply. In Chapter 10, we respond to some of the most common charges with some basic facts.
- The last two chapters concern how the

solar energy future we describe can become a reality. Chapter 11 provides an overview and assessment of various types of policies. Largely considered the best policy, feed-in tariffs are the focal point. But the long-term expansion of renewables will have to include additional instruments, such as for the heating sector. We also briefly present the history of the concept behind feed-in tariffs, which go back to the Aachen Model of 'cost-covering compensation'. Finally, in Chapter 12 we present a number of examples of creative marketing strategies that have successfully sped up the implementation of renewable energy (mainly in communities). In doing so, we hope to provide some ideas of how people and communities can become involved in addition to actions taken by big energy players.

Renewable Energy – The Facts has a special design: each page of text has a chart juxtaposed. The concept is intended to give readers a quick overview of the topic. At the same time, we as authors are forced to cover each issue on exactly one page. In some cases, some ancillary ideas had to be deleted and moved into footnotes. To facilitate readability, we have also added a glossary of technical terms. Interested readers will also want to consult the list of important publications and websites to help them keep up with current events and find additional information on special topics.

This book is a translation of the third edition of the German publication; some of the data in the German book were updated for the English publication.

We hope that you enjoy the English version of this book and find that it provides you with the basic knowledge you need to get involved in sustainable energy policy. There may be many setbacks to come, but one thing is also certain: the course of the sun cannot be stopped.

Dieter Seifried and Walter Witzel
Freiburg, March 2010

PS All the figures in this book can be downloaded at www.earthscan.co.uk/onlineresources. We hope they prove useful to you in your presentations and awareness-raising.

New Paths to the Future

Dear Readers,

In the battle against climate change, practical expertise in energy efficiency and renewables is in higher demand than ever. After all, renewable energy represents a truly long-term alternative compared to finite, environmentally unfriendly fossil energy sources – which are also unsafe in terms of security. The inexhaustible power of the sun is not the only way to fulfil our responsibility to future generations; wind, water and renewable bioenergy are of help and can be used as well.

Renewables offer genuine hope for development because they can provide decentralized energy in developing coun-

tries; therefore, they are used wherever poverty and a lack of energy would go hand-in-hand. They are also useful wherever people already have a lack of means to deal with the consequences of the wrong energy policy and environmental disasters such as droughts, floods and hurricanes.

Renewable Energy – The Facts provides a number of important answers to a lot of such urgent questions. It offers the latest information and technical explanations, including interesting examples and how to put guidance into practice. An agency of German development cooperation, InWEnt (Capacity Building International, Germany) supports this publication. The promotion of renewable energies and energy efficiency for developing countries is at the core of Germany's policies to combat climate change and to foster climate adaptation.

Climate and energy policy is not simply a matter for national governments. Politicians, even at the most local level, are also concerned as are the private sector and individuals. After all, energy consumption and climate change make themselves felt in individual homes and businesses. Roughly 75 per cent of energy consumption takes place in cities, which is why sustainable energy policy has to be implemented there. Furthermore, the avoidance of carbon emissions and climate adjustments has to focus on urban areas. Worldwide, megacities and metropolises have the greatest need for action. These are the places where climate change is caused – and where the changes are felt the most. In particular, the fast-growing Asian megacities are often located on rivers and coasts, where the rising sea level caused by climate change is not an abstract idea but an everyday reality – along with increasingly frequent typhoons and floods. The poor people in shanty towns with the least money will pay the highest price.

Cities are strong and flexible enough to implement a new energy policy that will take them in the right direction; national governments, in contrast, often have sluggish governmental procedures, and resolutions take time to be adopted. But thanks to their close contact with citizens and the private sector, city governments are more able to raise awareness and implement innovative policies.

Renewable Energy – The Facts contains a number of useful ideas easy to apply. It is a must-read for anyone who wants to act responsibly and take advantage of the opportunities which the future offers.

Luiz Ramalho
Director of the Department of Sustainable Economy
InWEnt

1 Introduction

1.1 Our climate is at stake

Climate change is already making itself felt. Over the last century, the average global temperature rose by 0.7°C. Glaciers in the Alps are retreating, as is the Arctic ice shelf. The frequency and strength of hurricanes has increased, and extreme weather events – such as Hurricane Katrina in 2005 and the heatwave in Europe in 2003 – are becoming more common.

The causes are well known. When fossil energy is burned, carbon dioxide (CO_2) is released. Its concentration is increasing in the atmosphere, strengthening the greenhouse effect. Since the pre-industrial age, the concentration of CO_2, the most important heat-trapping gas, has risen from roughly 280 parts per million (ppm) to the current level of almost 390ppm. But CO_2 is not the only heat-trapping gas emitted by civilization. For example, large amounts of methane are released by farm animals and in coal and natural gas extraction. Likewise, laughing gas (nitrous oxide, N_2O) is a heat-trapping gas from agricultural fertilizers.

These gases change the amount of energy trapped in the atmosphere and the amount reflected back into space. Shortwave sunlight penetrates the atmosphere and is reflected from the Earth's surface. Reflected waves are generally longer and cannot penetrate the atmosphere as well; heat-trapping gases partially absorb them. This natural phenomenon (the greenhouse effect) is vital for our planet; without this effect, the Earth would have an average temperature of −18°C. The increasing concentration of these heat-trapping gases is gradually disturbing this ecological equilib-rium. Land and oceans are heating up faster, more water vapour evaporates from the seas, and hurricanes and typhoons are becoming more common. The overall amount of energy input into the atmosphere is increasing. As a result, extreme weather events such as droughts, floods and heatwaves are becoming more common.

A decade ago, the idea that climate change was man-made was still controversial, but today there is a widespread consensus: 'Nowadays, no serious scientific publication disputes the threat that emissions of greenhouse gases from the burning of fossil fuels poses to the climate', says Professor Mojib Latif from the Leibniz Institute of Marine Sciences at the University of Kiel, Germany.[1]

Nonetheless, there is still some resistance to efficient climate protection policy, though this opposition is not the result of honest doubts about climate change. Rather, some industrial sectors simply have an eye on their bottom line and are concerned that their profits may suffer, as some countries and lobby groups would have us believe.

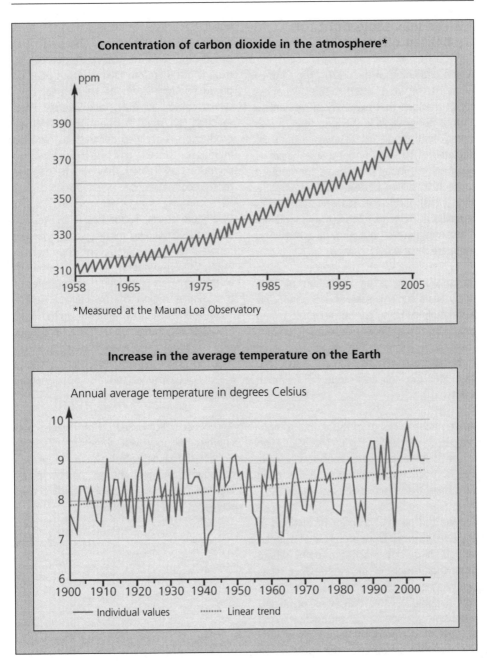

Figure 1.1 Our climate is at stake

Source: Al Gore, *An Inconvenient Truth*, 2006; BMU

1.2 The inevitable fight for limited oil reserves

At the beginning of the 1970s, the Club of Rome's *Limits to Growth* raised awareness about the idea that exponential growth on Earth is not possible in the long term. It also stated that crude oil reserves would be depleted in 30 years under a specific set of assumptions. Today, oil reserves are reported to be 1200 billion barrels (a barrel contains 159 litres), and the statistical range is reported as 42 years.[2] Those may sound like reassuring figures, but they are not. And there are several reasons why.

Statistical range is an indication of how many years current reserves – economically extractable oil using current technology and assuming that consumption remains constant – will last. But of course, if oil consumption continues to increase as in the past, then the statistical range will be much shorter.

While new sources of oil were found regularly up to the beginning of the 1980s, no major discoveries were reported in the 1990s. Since then, far more oil has been consumed than discovered (see Figure 1.2).

Our current oil fields cannot be drained at any rate we wish. Once an oil field has been tapped, it quickly reaches a point where production cannot be increased. Once it has been half emptied, one speaks of a 'depletion midpoint'. After that, it is practically impossible to speed up production. And because most current oil fields have already reached that midpoint, the production capacity of all oil fields in the world will begin to fall sooner or later – even though the range may statistically hold out for a few more decades. A number of oil-producing countries – such as the US, Mexico, Norway,

Egypt, Venezuela, Oman and the UK – have already passed their production peak, and others are soon to follow. A number of experts are therefore talking about peak oil production for the world – called 'peak oil' – which some say may have already been reached or may happen soon.[3] When production is likely to decrease as demand increases, prices can be expected to skyrocket, as indeed they did before the recent economic crisis.

One more crucial factor has to be kept in mind: the remaining oil reserves are largely found in a small number of countries. In 2005, OPEC members had three quarters of all proven reserves. Indeed, five countries of the volatile region of the Persian Gulf – Saudi Arabia, Iraq, Kuwait, the United Arab Emirates and Iran – alone make up 60 per cent of global oil reserves.[4] Instability therefore not only results from the absolute scarcity of oil reserves, but also from unequal distribution.

An energy policy based on renewables and energy efficiency will therefore not only protect the climate, but also make us less dependent on fossil energy, thereby reducing the potential for armed conflict over scarce reserves and resources.

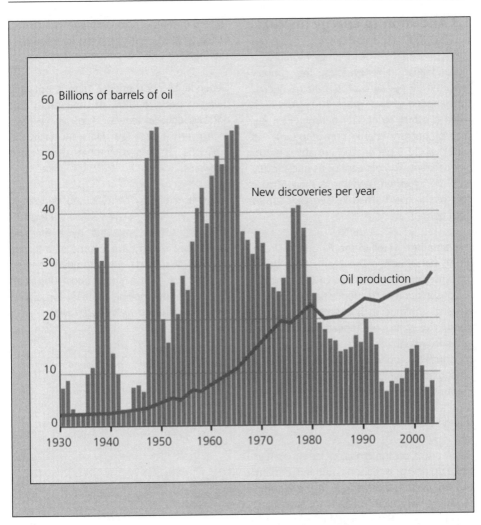

Figure 1.2 Oil reserves: The gap between new discoveries and production widens

Source: BP, IEA, Aspo, taken from SZ Wissen 1/2005

1.3 Addiction to energy imports

Though Germany is sometimes touted as a global leader in renewables, the country imported 59 per cent of its primary energy consumption as oil or gas in 2005. And even when it comes to nuclear energy (12.5 per cent of primary energy consumption) and hard coal (12.9 per cent), Germany is hardly independent; 100 per cent of its nuclear fuel rods are imported, and more than 50 per cent of the coal burned in Germany comes from abroad.

The situation overall in the European Union (EU) is hardly better. The 25 member states currently import around half of their energy. If consumption and domestic production were to continue in line with the current trend, the share of imports would soon exceed two thirds. Domestic production continues to drop within Europe, but energy consumption is increasing considerably. As a result, the share of domestic energy will continue to drop if energy policy fails to change these trends.

Rising prices on the global crude oil market woke up energy politicians both in Germany and the EU a few years ago. In the autumn of 2005, oil prices began to skyrocket, reaching prices that surprised many; a barrel of crude oil (159 litres) was being sold for more than US$70. But even that price would double before the economic crisis suddenly brought prices back down. The effects of this price hike made themselves felt in consumer prices. While a family that consumes 3000 litres of heating oil a year only had to pay around €1000 in Germany in 2003, that figure had doubled by 2005/2006 and would double again by 2008.

Oil and gas imports to Germany rose to €66 billion in 2005, a 27 per cent increase over the previous year.[5]

Dependence upon energy imports not only means a heavy outflow of capital, but also narrows political leeway[6] and, as we have seen over the past few years, increases the likelihood of armed combat over scarce resources.

Sustainable energy policy based on energy efficiency and renewables therefore strengthens local markets by redirecting capital that would have left the area to pay for energy imports into domestic energy sources. But such a policy also helps keep the peace by making battles for scarce resources unnecessary to begin with.

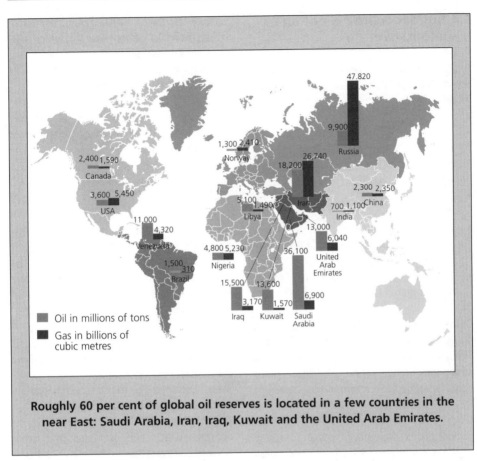

Roughly 60 per cent of global oil reserves is located in a few countries in the near East: Saudi Arabia, Iran, Iraq, Kuwait and the United Arab Emirates.

Figure 1.3 Global oil and gas reserves (2005) are restricted to a few regions

Source: BP Statistical Review 2006

1.4 Nuclear energy is not an alternative

A number of issues pertaining to nuclear energy have yet to be resolved and may be irresolvable:[7]

- The danger of a reactor meltdown like the one in Chernobyl (1986) remains, as events in July 2006 at Sweden's Forsmark nuclear plant revealed.[8]
- There is still no final repository for highly radioactive waste.
- The 'peaceful' use of nuclear energy cannot be completely separated from military applications.
- There is no perfect way to protect nuclear plants from terrorist attacks.

Germany, therefore, recently resolved to phase out its nuclear plants.[9] In 2005, the Obrigheim plant was the first to be decommissioned. Since then, nuclear plant operators have been attempting to overturn the agreement they themselves signed in order to have longer commissions for their nuclear plants. They have discovered a new argument: climate protection. They claim that nuclear power would have to be replaced by coal plants and gas turbines, which produce more CO_2 emissions than nuclear plants, thereby running contrary to current efforts to reduce these emissions.

Nonetheless, longer commissions for nuclear plants are the wrong way to get out of our climate trap, as would be newly constructed nuclear power plants. For instance, if we want to use nuclear power to ensure that we reach the German goal of an 80 per cent reduction in CO_2 emissions below the level of 1990 by 2050, Germany would have to construct and operate some 60–80 new nuclear plants, roughly 4–5 times more than the current 17 nuclear plants in Germany.[10]

Globally, several hundred new nuclear plants would need to go on line to reduce CO_2 emissions considerably; some 440 are currently in operation. In turn, nuclear risks would increase significantly.

At the same time, the supply of nuclear fuel rods is hardly ensured. At current rates of consumption, uranium reserves will only last for another 40–65 years.[11] If we build new nuclear plants, the uranium would not even last that long.

In addition, investments made in the energy sector clearly revealed that nuclear power is now considered too expensive. Since the disaster at Chernobyl, very few new nuclear plants have been ordered, and the ones that were built generally received generous subsidies.[12]

The nuclear industry would have us believe that nuclear power is undergoing a renaissance. Lobbyists like to point out that a few plants are currently under construction, but those figures include discontinued projects abandoned years ago. And because so many nuclear plants will be decommissioned over the next few years worldwide even under the normal schedule, the number of nuclear plants will decrease.[13]

At the beginning of 2007, for example, seven nuclear plants in Europe were taken off the grid for good – four of them in the UK, two in Bulgaria and one in Slovakia.[14] The risks of nuclear power can be prevented if we switch to renewables, which are an environmentally friendly alternative (see 11.18).

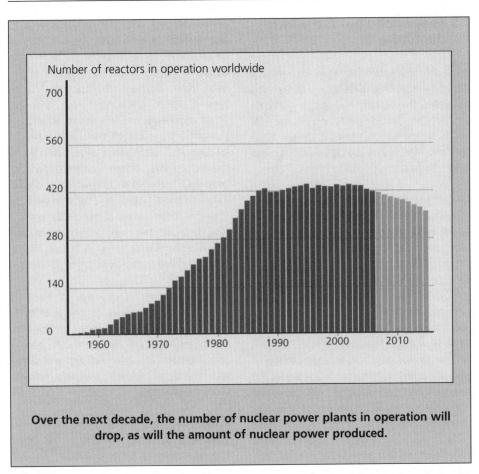

Number of reactors in operation worldwide

Over the next decade, the number of nuclear power plants in operation will drop, as will the amount of nuclear power produced.

Figure 1.4 Nuclear power is not an option

Source: IAEA

1.5 Renewables are the way of the future

While the share of renewables in Germany has been increasing drastically over the past ten years, the country still gets around 85 per cent of its energy from fossil sources. Oil makes up the largest share of the pie at 36 per cent, followed by coal at 24 per cent and gas at 23 per cent.[15]

Up until the 18th century, civilization got all of its energy from such renewable sources as wood, wind, water and muscle. Coal – and later oil and gas – only took off at the beginning of industrialization. Today, we admittedly do not face any acute shortage of fossil energy, but reserves are nonetheless finite. Estimates are that, under current consumption, known reserves of oil will be depleted in some 40 years and brown coal in 220 years.[16] And while new resources may yet be discovered, these resources remain finite. Figure 1.5 clearly shows that the age of fossil energy will only appear as a blip on the screen of energy consumption over a 4000 year period.

When coal, oil and gas are combusted, CO_2 is released. CO_2 is the main reason why the Earth's atmosphere is heating up – and why the climate is in danger. Back in 1990, the German Parliament's Commission on Protecting the Earth therefore called for an 80 per cent reduction in CO_2 emissions within Germany by 2050.

Nuclear power currently covers some 30 per cent of electricity consumption in Germany, roughly 13 per cent of the country's total energy consumption. However, the risk of a major reactor meltdown like the one in Chernobyl in 1986 and the unsolved problem of waste disposal rules out this high-risk technology as part of a sustainable energy supply.

So we are left with one major source of energy over the long term: the sun.[17] It can be used both to generate heat and electricity (solar thermal and solar power). At the same time, however, the sun is also the reason plants (biomass) grow and the weather changes. Different amounts of sunlight hit the Earth's surface in different locations, bringing about wind.[18] And when the sun shines, ocean water evaporates, creating clouds and rain – hydropower. Biomass, wind power and hydropower are therefore indirect ways of using solar energy. And because the sun, wind, water and biomass are inexhaustible in human terms, they are called 'renewable' types of energy.

There are a number of ways to use solar energy directly and indirectly, and all of them are constantly being further developed. Along with greater efficiency in energy consumption (see 1.8), indirect and direct solar energy will provide a reliable, sustainable energy supply.

Renewables are the way of the future. The Solar Age will arrive one way or the other. The question is only whether we will manage that transition without crises and major conflicts.

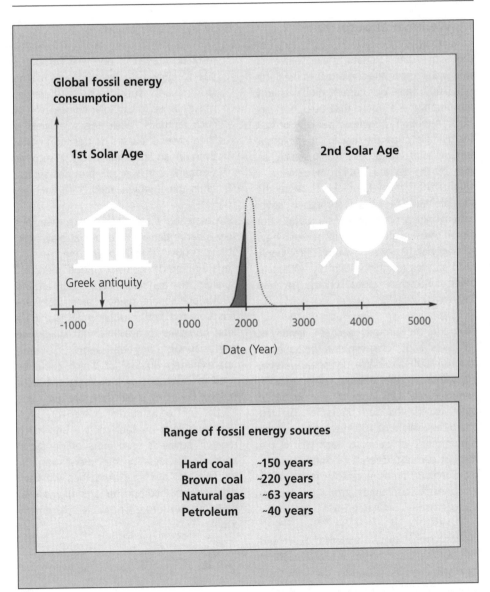

Figure 1.5 The sun is the future

Source: Goetzberger and Wittwer, Sonnenergie; Bundesamt für Geowissenschaften und Rohstoffe, 2006

1.6 We have enough sun

The sun is our planet's main source of energy. Each year, the sun provides the Earth with 7000 times our current global energy consumption – a figure that does not vary much. Although roughly 70 per cent of that energy falls onto the ocean, there is still enough solar energy left. For instance, an area of the Sahara 200km by 200km – roughly the size of Kentucky or twice the size of Wales – would suffice to cover current global energy consumption. But even if this sunlight could only be used at an efficiency of 10 per cent, we would still only need an area roughly 700km by 700km to cover our current global energy demand with solar power.[19]

Of course, the sun only reaches Germany at half the strength of sunlight in the Sahara – roughly 1000–1100kWh per square metre, equivalent to the amount of energy in approximately 100 litres of heating oil. In other words, the sun pours roughly the energy equivalent of 100 litres of oil on each square metre of Germany each year in the form of sunlight. Overall, Germany receives more than 80 times more solar energy than it currently consumes from all energy sources.

Sunlight comes in two varieties: direct and diffuse. The latter occurs when sunlight is reflected, such as in clouds. The light then reaches the surface from various directions. Some solar energy systems need direct sunlight (see 2.6), but most can utilize both direct and diffuse sunlight.

These figures clearly show:

1   Insolation, even in northern Europe, is still roughly half as strong as in the tropics and subtropics. It therefore makes sense to use solar energy even at such latitudes. While the solar yield is then lower, there are no transport costs.
2   Even in an industrial country such as Germany, the sun still provides several times the energy needed.

The benefits of solar energy are clear, but low 'energy density' is a crucial drawback. While 1000W of solar power may reach a square metre of northern Europe under full sunlight, the annual average is only around 100W per square metre. Large areas are therefore required for solar arrays. But if we limit ourselves to available roof space, we see that Germany has some 3500km², approximately 800km² of which could be used for solar energy.[20] With current technology, Germany could therefore get some 16 per cent of its current power consumption from solar on such roofs – and much more if power is used more efficiently. In addition, facades, bridges, noise barriers, etc. are also available. And then we have wind power, hydropower and biomass to round off our future renewable solar energy supply.

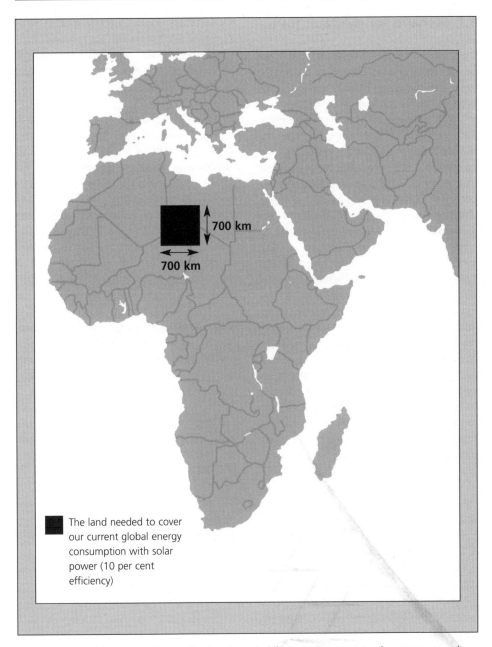

Figure 1.6 We have enough sun: The land needed for a 100 per cent solar energy supply

Source: The authors

1

1.7. Scenario for the solar future

If we are to change our economy so that we can get most, and possibly all, of our energy from the sun and other renewables, we need to change our energy policy first. Back in 1980, the Institute of Applied Ecology in Freiburg, Germany, worked up a scenario for this transition.[21] The main thing that we have to change is our minds: the focus does not need to be on greater energy consumption, but on greater prosperity. Entitled 'Energiewende' (Energy Transition), the Institute's study therefore took a look at society's needs for energy services, such as lighting, transportation and heated buildings. The energy required for these tasks not only depends upon the scope of these services, but also on energy efficiency. If, for instance, gas mileage can be tripled, people could then drive three times as far with the same amount of energy – or 50 per cent further with half as much fuel. The study demonstrated such efficiency potential in a number of fields. It concludes that we can reduce our primary energy consumption by nearly 50 per cent over the next 40–50 years even as our standard of living continues to increase.

These findings have been confirmed again and again since:

- The 11th German Parliament's Commission on Protecting the Earth's Atmosphere found that energy savings of 35–40 per cent are feasible.[22]
- In *Factor Four*,[23] Ulrich von Weizsäcker, Amory Lovins and Hunter Lovins tell the Club of Rome that the efficiency gains are so great that standards of living could be doubled even as energy consumption is cut in half.

- Another study in Germany, entitled 'Lead Study 2007 – Update and reassessment of the use of renewable energies in Germany' showed the potential and costs of this transition.[24]

In addition to demonstrating the great savings potential, the Energy Transition study conducted by the Institute of Applied Ecology also includes a scenario for the solar future. For example, solar energy can provide low-temperature heat. A greater share of wind and hydropower would cover our electricity consumption. Waste from the timber and agricultural sectors would provide heat, electricity and motor fuels. If the conservation potential is fully exploited, our energy supply could be redesigned so that solar, wind, hydropower and biomass cover roughly half of our energy consumption by 2030. The other half would then mainly come from coal in highly efficient, and therefore environmentally friendly, cogeneration plants.

The Energy Transition study does not specifically talk about a solar economy as a goal, but it does emphasize three important steps on the path towards a solar economy:

1 Conservation should be exploited whenever possible.
2 Cogeneration should be used as a bridge technology.
3 Renewables must be expanded.

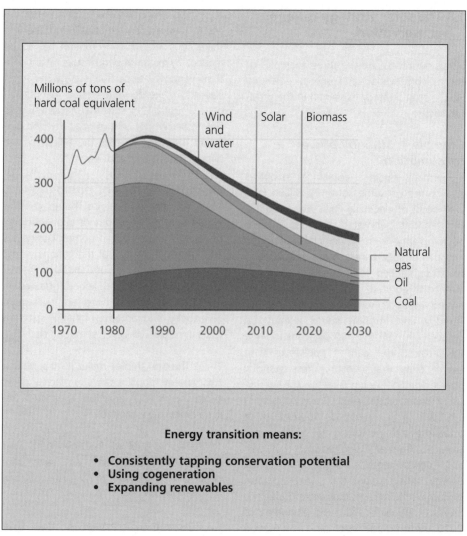

Figure 1.7 The Institute of Applied Ecology's Energy Transaction study (1980): Growth and prosperity without oil and uranium

Source: The authors

1.8 The solar strategy requires conservation

Let us now focus on the three examples of energy conservation and energy efficiency for the solar strategy explained in the previous section.

Example 1: Standby power consumption

In Germany, electrical appliances in offices and homes consume some 22 billion kilowatt-hours of electricity each year,[25] roughly the amount generated by four nuclear power plants. If this electricity had to be provided by solar panels, more than 200km^2 would be needed. In light of the costs and materials required, the effort would be absurd, especially when we realize that this standby consumption could already be reduced by more than 80 per cent today if we replaced our current appliances with newer ones that consume less standby power. The Eco-design Directive for Energy-using Products (2005/32/EC) was adopted in 2005 and came into force in August 2007. It establishes a framework under which manufacturers of energy-using products (EuP) will, at the design stage, be obliged to reduce the energy consumption and other negative environmental impacts that occur during the product's life cycle. From the beginning of 2010, the 'off mode' electricity consumption of all appliances sold in Europe is not allowed to exceed 1W and the stand-by mode is limited to 2W.

And if we switch appliances off completely (i.e., do without standby mode), we can reduce our consumption even further.

Example 2: Space heating

The average German single-family detached house with 120m^2 of floor space consumes some 30,000kWh per year for heating and hot water. A large solar hot water system (12m^2) can produce some 13 per cent of the energy required for that task. To increase that share considerably, consumption has to be reduced by means of good insulation and efficient windows. Such 'low-energy buildings' (see 4.1) make do with around 10,600kWh per year. But a 12m^2 solar thermal array would then cover 28 per cent of peak demand (see Figure 1.8). The greater the efficiency, the greater the share of solar energy.[26]

Example 3: Efficient pumps

More efficient pumps and pump controls would save many billions of kilowatt-hours of electricity and heat in homes, businesses and industrial plants. But this change would require decision-makers to be better informed and tradespeople better trained; in addition, we would need an investment philosophy that accepts higher investment costs in return for lower operating costs.[27]

These three examples make it clear that a solar energy supply is easier to reach and less expensive if conservation measures are simultaneously exploited.

Towards the goal of 100 per cent solar energy, Example 2 does not seem that convincing. If we want to go further, we can do the following:

- Use more solar energy. A larger solar array would cover a larger share of heating demand (see 2.3).
- Use greater efficiency. By further improving insulation and ventilation systems, we can do without heaters altogether (see 4.6).
- Use other types of renewable energy. A combination of solar collectors and wood heating can provide renewable heat all year round (see 4.8 and 5.4).

These options represent a good starting point for the transition to the Solar Age.

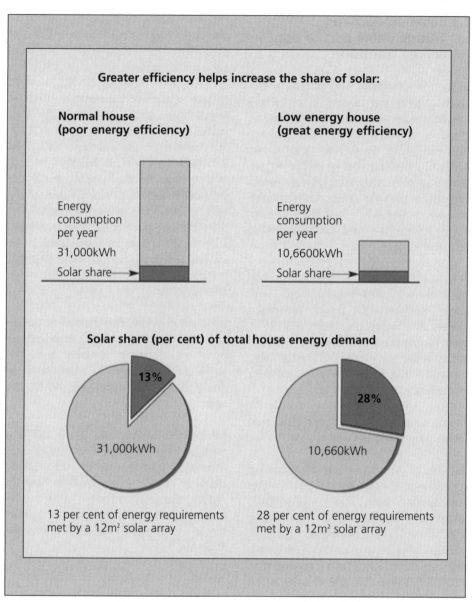

Figure 1.8 Solar strategy requires energy conservation

Source: The authors

1.9 Cogeneration – an indispensable part of our energy transition

Today, the generation of electricity is the cause of more than a third of all carbon emissions in Germany. The reason is the low efficiency at which power plants convert fuels into power. On average, fossil power plants run at efficiencies far below 40 per cent. If we then deduct the power needed by the plant itself and transport losses on the grid, we see that only a third of the primary energy fed into the plant actually arrives at your wall socket.[28]

The alternative to conventional power generation is called cogeneration. Here, waste heat from the power generation process in conventional steam turbines is used. For the waste heat to be used in residential areas, hospitals or commercial units, the power has to be generated close to consumers.

The overall efficiency of cogeneration units ranges from 85–95 per cent.

Because of this high rate of efficiency, cogeneration units are not only much better ecologically, but also economically. Nonetheless, cogeneration plants make up less than 10 per cent of installed capacity in Germany because large utilities have consistently attempted to stamp out cogeneration efforts by communities and industry, which would have cut into the sales revenue of utilities.[29]

The liberalization of the power market made the efforts to stamp out cogeneration even fiercer. Large power producers, all of whom suffered from overcapacity, lowered their prices to cutthroat rates so that the already installed fleet of cogeneration units was no longer profitable. With prices at 2–3 eurocents per kilowatt-hour – below the cost of production – even highly efficient cogeneration cannot compete.

To take account of the negative effects of liberalization on cogeneration, Germany passed its Cogeneration Act in March 2002. The goal was to reduce carbon emissions by 23 million tons annually by 2010 through cogeneration. In all likelihood, this target will not be reached. One of the goals of the new governing coalition in Germany is therefore to respond to calls by the cogeneration sector and improve legislation.[30] The UK has also come up with some proposals for proper compensation of heat in its Energy Bill.

We need only look elsewhere in Europe to see how effectively cogeneration can be used. In Denmark, Finland and The Netherlands, the share of cogeneration in power production is between 40–50 per cent.[31]

For the next few decades, we will still have fossil power plants generating electricity. It is therefore crucial that we use these power plants as efficiently as possible in order to reduce the environmental impact. Like energy conservation, cogeneration is therefore a crucial part of our transition to the Solar Age.

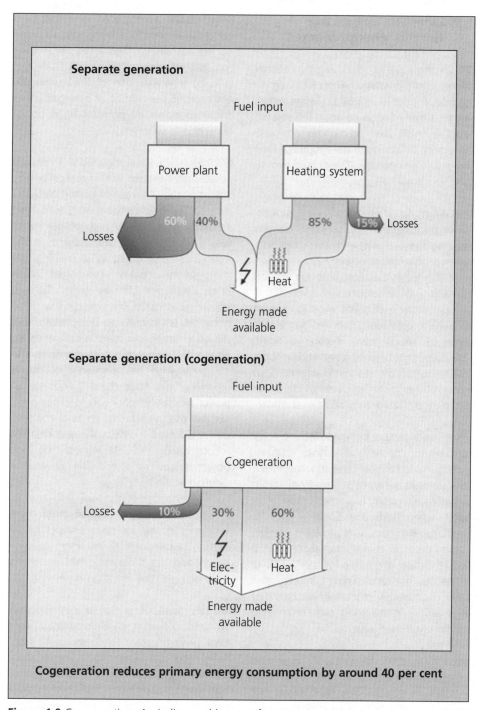

Separate generation

Fuel input

Power plant

Heating system

60% 40%

85% 15% Losses

Losses

Heat

Energy made available

Separate generation (cogeneration)

Fuel input

Cogeneration

Losses 10%

30% 60%

Electricity

Heat

Energy made available

Cogeneration reduces primary energy consumption by around 40 per cent

Figure 1.9 Cogeneration: An indispensable part of our energy tradition

Source: The authors

1.10 Liberalization of the German energy market

On 29 April 1998, the German energy market was 'liberalized'. From one day to the next, former monopoly power providers saw their markets opened up to the competition; after 60 years of monopoly service, customers could now (temporarily) look forward to competition. They could choose their own power provider.

The effects of liberalization were drastic. Power prices initially fell, providers merged and the big fish acquired the small fish. Power providers have increasingly focused on what they see as their core business: increasing sales revenue. In particular, overcapacity and predatory pricing put more and more pressure on community cogeneration plants, some of which were decommissioned. Drawn up at the beginning and middle of the 1990s, least cost planning schemes[32] to increase the efficiency of electricity consumption were put on ice or discontinued.

Competition temporarily made it cheaper for families to consume electricity. Initially, experts expected retail rates to remain basically stable in the wake of liberalization, but something surprising happened in the fall of 1999, when RWE and EnBW – two of Germany's Big Four power providers – began cutting prices to gain market share. Only a few years later, the battle for a larger share of the retail market died down. Indeed, retail rates are currently much higher than they were before liberalization and continue to rise far faster than inflation.[33]

From 2002–2007, for example, retail rates rose by around a third without any increase in taxes on power.[34] The main reason for these price hikes is the market power of the Big Four and the lack of competition. After a brief phase of fierce competition (1998 to

2000), E.ON, EnBW, RWE and Vattenfall realized that the best strategy was to divide up the pie among themselves rather than compete for a bigger slice. Their strategy is working quite well; after all, the four oligopolists run 96 per cent of all baseload plants and account for 80 per cent of all power generated in Germany.[35]

In 2005, politicians responded to increasing power prices and the lack of competition by creating the German Federal Network Agency. The Agency has already succeeded in lowering excessively high power transit fees in a number of cases. But even the Agency can only go so far in creating true competition between companies. Alois Rhiel, Economics Minister in the State of Hessen, thus called for antitrust law to be made stricter in order to demonstrate that the government can make a difference. As he put it: 'Otherwise, the state will have to do away with the oligopoly of power producers and force the Big Four to sell power plants.' His goal was to increase the number of power producers until competition could get a foothold, the goal being to reduce retail rates. He argued that high power prices were bringing down the German economy.[36]

In October 2006, the German government took another step to ensure competition by making it easier for retail customers to switch power and gas providers. Now, customers need only give one month's notice.

At this point, it is up to customers to demand competition. Unfortunately, they have been reluctant to do so up to now. From 1998 (the beginning of liberalization) to the end of 2006, fewer than 5 per cent of household customers switched power providers even though they could have saved a lot of money by moving to a provider with lower rates.[37]

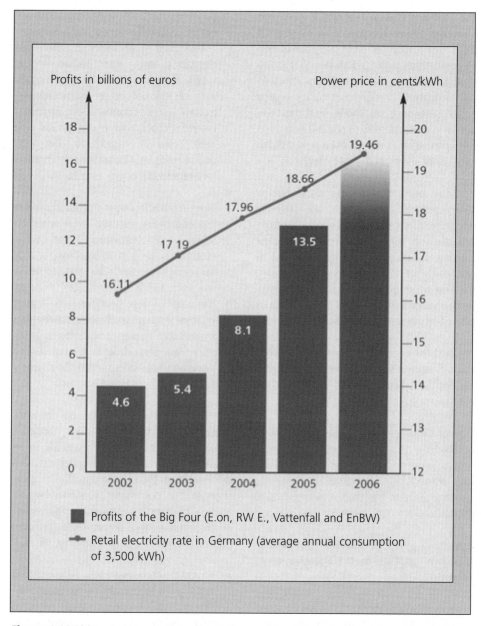

Figure 1.10 Rising power prices: Profits at the expense of households and small consumers

Source: The authors

1

The changing market has also endangered municipal power companies. A number of them have responded proactively out of fear of competition and sold shares to the Big Four. In doing so, they generally also sold their leeway to design community energy policy. Experience has shown that the business and sales interests of the Big Four then take precedent, even if they only hold a minority of shares in the municipals.

Yet, the prospects for municipal energy service providers are quite good.[38] On the one hand, they can pick and choose a power provider; on the other, they know their customers well and have the expertise to offer the kinds of energy services in demand. But they will only be able to justify their existence if their services differ from the ones offered by firms motivated solely by profits.

The next few years will show whether municipal utilities will succeed even if their prices are not the lowest provided if they offer the energy services customers demand.[39]

One of those services is a commitment to renewables. But merely providing 'green power' is not enough; rather, municipal utilities should help construct and finance systems owned by their customers and promote community projects.

Two main pillars of the energy transition have been crippled by the liberalization of energy markets:

1 Power generation in municipal cogeneration units and industry has been ramped down. Because the Big Four have used their overcapacity to pursue predatory pricing, cogeneration no longer pays for itself. The German government reacted to the inevitable with a special law to promote cogeneration, but the goal of reducing emissions

by 23 million tons of CO_2 by 2010 will not be reached under current legislation.

Improved legislation is therefore needed quickly; after all, an energy supply based on solar energy cannot do without efficient power plant capacity, exactly what cogeneration provides. Indeed, cogeneration plants can not only help cover the baseload, but also compensate for fluctuations in intermittent renewable power (see 10.1).

2 Since the beginning of liberalization, the Big Four have focused even more on sales of kilowatt-hours. Yet, what customers need is inexpensive energy services. After some success stories in the mid-1990s, energy conservation campaigns have become rare among utilities. And while customers benefited from lower power prices initially, today they can only defend themselves by switching providers, which they largely fail to do.[40]

Our experience with liberalization demonstrates that the market has to be changed if the environment and the climate are to be protected. The market needs guidelines so it will develop in the desired direction. If these changes are not made, liberalization will have a tremendous environmental impact and be a tremendous burden on the German economy. So where do we go from here?

• Our tax system has to be systematically redesigned to increase the tax burden on energy consumption and resource consumption and lessen the burden on labour. What we need is the kind of environmental tax reform that Germany began ten years ago with its 'eco-tax' on fossil fuels. But the current tax rates on energy consumption and private homes, businesses and industrial plants are not yet high enough to ensure that energy is consumed efficiently. Further steps are necessary.[41]

- We need a tradable quota for cogeneration. Utilities should be required to cover a certain share of the power they sell with cogeneration units. The utilities could then decide whether they want to generate that power themselves or purchase it. Current German law, based on a floor price (feed-in tariff) for kilowatts of electricity exported to the grid, cannot ensure the success of cogeneration as well.

- A special fund[42] for energy conservation and/or energy efficiency should be created, and further steps should be taken to promote efficient technologies (demonstration projects, training programmes, better labelling on large electronic devices, energy ratings for buildings, stricter efficiency standards, etc.). An energy efficiency fund would publish competitive requests for energy conservation proposals. The company that can provide the required energy conservation at the lowest cost would be awarded the contract. Such energy efficiency funds have already proven successful in a number of countries.[43] The efficiency fund could be financed by means of a surcharge on the retail electricity rate or on the other energy carriers.

- Stricter energy conservation requirements for new buildings and renovation projects.

- Utilities must be forced to compete for customers to a greater extent so that utilities will change the services they offer. The price per kilowatt-hour should not be the only reason to switch utilities. Rather, customers should also take other economic, ecological and social aspects into account. To this end, the performance of utilities should be rated. Such ratings would provide benefits not only to customers, but also to companies

with an eye on long-term success.[44]

- New regulations are needed to allow grid operators to pass on the cost of investments in energy conservation campaigns to their customers, which they are currently not able to do.

- Renewable energy sources should receive assistance for market launch, and feed-in tariffs must be offered for renewable power exported to the grid. This is where governments can make the greatest difference (see 11.9–11.12).

- Nuclear power has to be phased out as quickly as possible (see 1.4 and 11.18) and energy policy has to be restructured accordingly. Here, it is crucial that the EU, member states and regional governments state specific targets and pursue a reliable energy and transport policy to provide a stable business environment for investors, manufacturers of efficiency technology and producers of renewables technology over the long term. A number of new market players will be required in the process: energy agencies, contracting firms,[45] and academic centres that offer degrees in energy. Companies focusing on efficiency technology could step up energy and electricity conservation. These firms will, however, only be successful if the business environment fosters the kind of work they do.

Clear signals set by reliable taxation and incentives for consumers are equally important. Indeed, they are indispensable if consumers are to take energy costs into greater consideration when purchasing items with long service lives, such as cars and refrigerators.

We already know that a sustainable energy supply is not a pipe dream, but a feasible alternative.[46] It is also affordable and would provide significant economic and social benefits.

1.11 Economic benefits

Considerable investments will have to be made for our energy supply system to become sustainable; renewables will have to make up a large share of our supply, and energy applications will have to become much more efficient. These upfront investments are offset over the long term by lower energy costs and independence from fossil energy sources, which are only going to become more expensive.

At current oil and gas prices, a wide range of investments in energy efficiency already pay for themselves and provide both macro and microeconomic benefits without any subsidies.[47] In contrast, renewable energy sources require further support, such as from feed-in tariffs, which ensure a return on investments in renewables (see 11.10). But the surcharges that feed-in tariffs entail do not simply increase the price of electricity for consumers; rather, the growing share of renewable energy sources has brought power prices down on power exchanges, as recent independent studies have demonstrated.[48] While this may seem surprising at first glance, it makes complete sense. Prices on power exchanges are sorted in ascending order. The greater the demand for power, the more power plants with low efficiency rates and higher fuel costs are used. But if a large amount of renewable power is being generated, the worst power plants are no longer used. Supply and demand converge to produce lower exchange prices. On average, power prices in Germany have fallen as a result by as much as 0.76 euro-cents per kilowatt-hour.[49]

In particular, such macroeconomic benefits are the main reason why we should restructure our energy supply:

- External costs in our current energy supply – from air pollution, climate change, the risk of nuclear disaster and the cost of securing access to raw materials – are by definition not included in monthly power bills. Instead, they are incurred outside of our energy supply as external costs that society must cover. A sustainable energy supply largely avoids such external costs. Furthermore, there is the increasing damage caused by extreme weather. These costs have risen to nearly €2 billion per year on average for Germany alone, but the German Institute for Economic Research (DIW) estimates that that figure could rise to €27 billion by 2050.[50]
- A sustainable energy supply would create hundreds of thousands of jobs. In the renewables sector alone, 280,000 new jobs had been created by 2008 in Germany, and that figure could rise to 500,000 by 2020 (see http://bee-ev.de/Energieversorgung/Wirtschaftlichkeit/Arbeitsplaetze.php). Every new job is not only valuable from the social standpoint, but it also reduces expenditures for unemployment benefits.

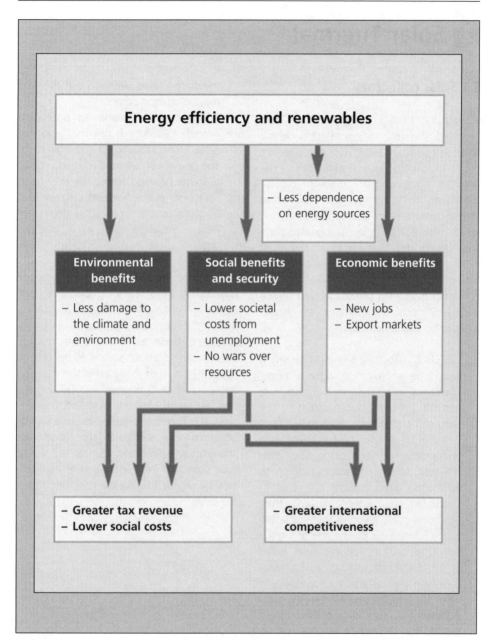

Figure 1.11 The economic benefits of energy efficiency and renewables

Source: The authors

2 Solar Thermal

2.1 Solar collectors

When sunlight hits a dark object, that object heats up. This well known effect is behind solar thermal systems. Solar collectors convert sunlight into usable heat. The absorber – a black plate made of copper, aluminium or sometimes even plastic in simple systems – is exposed to the sun, which heats the plate up. The heated plate then passes the heat on to a fluid (heat carrier) flowing through tubes embedded in the plate, and this fluid flows to the consumer device (to provide heated service water). Various technologies are imple-mented to reduce heat losses:[1]

- Flat-plate collectors, the most common type, have an absorber within a frame. The cover facing the sun is transparent, and the cover is well insulated to the sides and on the back. Such solar collec-tors are 50–60 per cent efficient at a temperature of 50° Celsius. They can, however, reach temperatures of up to 80°C under direct sunlight.
- Evacuated tube collectors have their absorbers inside a sort of thermos bottle that is transparent on the side facing the sun. The vacuum reduces heat losses greatly. These evacuated collectors there-fore have greater efficiencies, especially when the temperature difference is great between the absorber and the ambient air. If the difference is 70K (= 70 degrees), evacuated tube collectors are around 15 per cent more efficient than flat-plate collectors. But that figure drops to only 5 per cent if the difference is 40K. The benefits of evacuated tube collectors thus make themselves felt especially during the cold season. It therefore makes sense to use them to support heating systems.
- When outdoor swimming pools are heated, heat loss is not that great an issue, so no insulation is generally used. The water is pumped through black absorber hoses to absorb the solar heat. Such systems are, however, only useful in applications where only a few degrees of heat is required, such as in outdoor pools.

Solar collectors also have to be properly oriented to be efficient. In Germany, roof angles are optimally 45 degrees and facing due south. Fortunately, efficiencies drop only slightly if these ideal values are missed; collectors still produce around 90 per cent of their rated output if the collectors have an angle between 0 degrees and 50 degrees and face anywhere from southeast to south-west. It is therefore possible to use a slightly larger collector surface to compensate for suboptimal orientation. Some evacuated tube collectors even allow you to compen-sate by slightly turning the absorber within the tube.

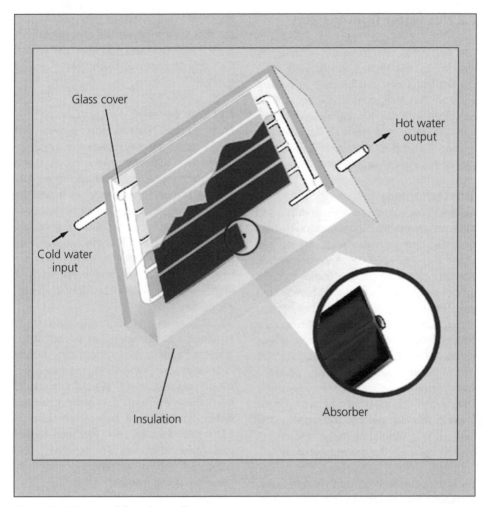

Figure 2.1 Design of flat-plate collectors

Source: Schüle, Ufheil and Neumann: Thermische Solaranlagen, 1997

2.2 Hot water from the sun

The heat from solar collectors is generally used to provide heated service water, but the hot water can also be used in heating systems. The solar thermal system is then combined with the heating system; generally, the solar heat suffices to cover all heating demand in the warm season and part of heating demand in the cold season.

The technology

Systems for single-family homes and duplexes are the most common today. They generally have a dual circuit system (see Figure 2.2). The solar circuit (solar collector, bottom heat exchanger and pump) contains a frost-proof fluid (a mixture of water and glycol) that transports heat to the heating circuit. Once the sun starts shining, the temperature in the collector rises several degrees above the temperature in the lower part of the tank, setting off the pump in the solar circuit by means of an electronic control. The solar heat then passes from the lower heat exchanger into the service water tank, which should be large enough to provide hot water for around two days.

Because hot water is lighter than cold water, the water heated with solar energy rises in the tank. The water thus separates into layers, with the lighter hot water at the top and the colder water at the bottom. Hot water is then drawn from the top of the tank, where the stored water is always the hottest. If the sun ever fails to provide enough heat, the auxiliary heater can provide extra heat via the top exchanger.

Example

A solar thermal system with some 6m² of flat-plate collectors and a 300 litre storage tank can generally cover hot water demand during the warm season in a four-person household. But even the solar heat during the cold season offsets the consumption of around 300 litres of heating oil per year in Germany. The net investment cost of around €5000.[2]

Decisive benefit

Solar thermal systems that provide hot service water not only replace the consumption of fossil fuels with solar energy, but also very effectively reduce pollution. Because oil and gas heaters run at low levels during the warm season, they are especially inefficient then and emit more pollution; the heaters generally only come on briefly and therefore do not run in their optimal range. But if solar heat is used during the summer to heat service water, you can leave your boiler off altogether.

In addition to providing hot service water, larger solar thermal arrays can also provide heat for space heating during the spring and autumn. In such cases, the building should be properly insulated and have a heating system running at a low temperature. A 12m² solar thermal array can then reduce fuel consumption by 20 per cent. This combined use is becoming increasingly common on the market.[3]

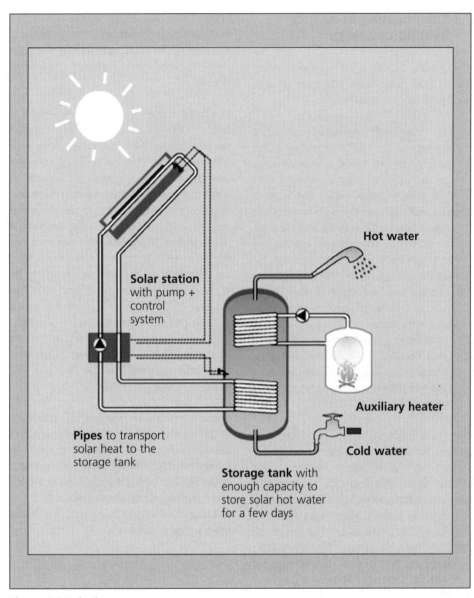

Hot water

Solar station
with pump +
control
system

Pipes to transport
solar heat to the
storage tank

Auxiliary heater

Cold water

Storage tank with
enough capacity to
store solar hot water
for a few days

Figure 2.2 Solar hot water

Source: Schüle, Ufheil and Neumann: Thermische Solaranlagen, 1997

2.3 Solar heating in district heating networks

If you want to get a lot of your heating energy from solar heat, you have to take two factors into consideration:

1   Far more heat is required for space heating than to heat service water. In old buildings, 5–10 times as much energy can be required for space heating than to provide hot tap water. But even in well insulated homes, space heating can require two to three times more energy. For solar heat, the first goal is therefore to reduce the building's requirements for heating energy; in addition, far greater collector areas are required when space heating is to be covered in addition to hot service water.
2   In the winter, when you need the most space heating, you have the least solar energy. In January, the sun only provides a sixth of the energy it supplies in July at Northern European latitudes.

The main problem with solar heating is getting the excess heat from the summer stored for the winter. Long-term heat storage tanks play a crucial role. Specific heat losses and specific costs both decrease as the size of these storage tanks increases. Therefore, subterranean storage tanks are generally the best option.[4] District heating networks can then provide heat to adjacent buildings from these underground tanks.

In other words, more solar heat can be provided when buildings are well insulated, collector fields are large, a long-term storage tank is available and a district heating network can be connected to it.

In the 1980s, Sweden set up the first district heating networks with solar heat. Germany followed suit in 1996, with its first two solar pilot systems that included long-term storage.

In a pilot project in Hamburg-Bramfeld, a total of 3000m^2 of collectors was installed on the roofs of 124 row houses in a new neighbourhood. The solar heat collected was stored centrally and distributed to consumers via a district heating network. Any heat not used was fed to the long-term storage tank, a 4500m^3 concrete tank with a stainless steel lining. The tank was insulated on the outside to reduce heat loss. It filled up with heat in the summer, reaching its highest temperature of around 80°C in August. It then provided heat for space heating well into December. A condensation boiler was used to provide heat for the rest of winter. The goal was to cover 50 per cent of energy demand for heating and hot water, and that goal was reached.

The cost of this solar heat (without subsidies) in the Hamburg project is 26 euro cents per kilowatt-hour, but another project in Friedrichshafen, which was larger and had a larger storage tank, provided heat at a price of 16 euro cents per kilowatt-hour. By 2005, five other solar district heating networks had been set up in Germany.[5]

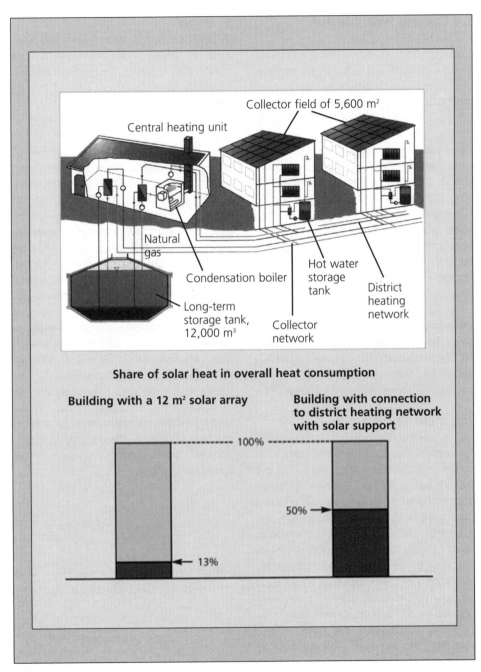

Figure 2.3 Solar thermal with long-term heat storage

Source: BINE, Solar Nahwärmekonzepte

2.4 Cooling with the sun

At present, most air-conditioning and refrigeration devices are vapour-compression machines. The refrigerants used, even though they do not contain CFCs, are not exactly environmentally friendly. And, of course, these units consume a lot of power. Here, solar energy can also help offset the consumption of fossil energy, reduce pollution and reduce peak power demand. It turns out that the most solar power is available right when air-conditioning is needed most. Buildings are not only mainly air-conditioned in the summer, but also during the day, which makes the costly long-term storage required for solar heat unnecessary here.

There are various types of solar cooling based on solar thermal heat.[6] Those based on low-temperature heat are especially useful for solar applications. Below, we discuss Desiccant Evaporative Cooling (DEC).

The technology

DEC is based on the principle that evaporating water cools down its surroundings.[7] A humidifier cools down the air by taking out the heat as water evaporates from the previously dried air.[8] The dehumidifier contains silica gel that removes moisture from the incoming air by absorbing the moisture in its own molecular structure. The silica gel therefore has to be 'regenerated' (dried) for reuse, so when the outgoing air passes by, it takes some of the moisture with it. The heat required in the process comes from solar collectors, but it could also come from a short-term storage tank with an auxiliary heater.

Example

In 1996, a solar DEC system went into operation as an air-conditioning system at a startup park in Saxony, Germany. Its 20m² of collectors cover 75–80 per cent of the air-conditioning required.[9]

Outlook

There have already been a number of successful demonstration projects pertaining to solar air-conditioning. Serial production is getting underway, but solar cooling still cannot compete at current energy prices. Cost reductions are expected,[10] and less expensive air collectors (see 2.5) are expected to help bring down prices further. Such air collectors are especially useful towards the equator. Properly planned buildings in central Europe generally do not need air-conditioning, though special-purpose rooms – such as for technology or assemblies – are an exception. Here, it would make sense to have the collectors used all year round. In the summer, they could provide air-conditioning; in the winter, space heating. In tropical countries, solar cooling can help store food. Clearly, solar air-conditioning and cooling systems have considerable potential.

Adsorption air-conditioning

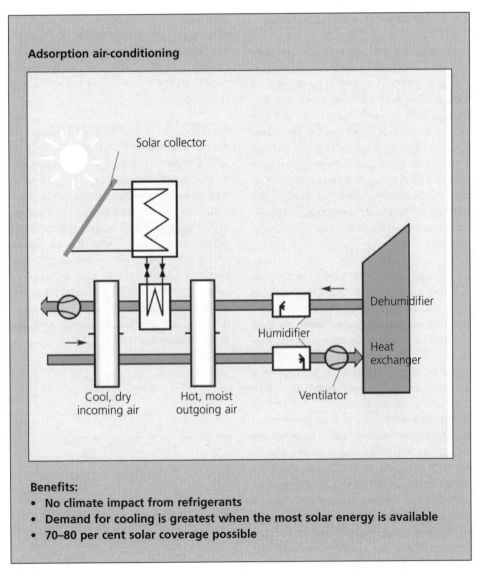

Benefits:
- **No climate impact from refrigerants**
- **Demand for cooling is greatest when the most solar energy is available**
- **70–80 per cent solar coverage possible**

Figure 2.4 Cooling with the sun

Source: Sonnenenergie 1/99

2.5 Solar drying – air collectors

Solar collectors containing fluids are fairly low-tech, but air collectors allow solar energy to be used for heating purposes in an even simpler way. Here, a sheet that lets through light is pulled taut across a frame just 2.5cm or so above a black plate. When sunlight hits this air collector, the air between the plate and the foil heats up. If the collector is set up at an angle, the hot air rises, automatically creating air flow. If the collector is completely horizontal, a ventilator is required, though it could run on solar power.

Air collectors allow solar energy to be used in drying applications. In many rural areas in the tropics and subtropics, farmers only have the wind and the sun to help them dry their harvests to conserve produce. Often, they simply spread fruits and vegetables out under an open sky. Some of the harvest is generally lost to rodents and birds, while dust, microorganisms, mildew, etc. reduce quality. As a result, part of the harvest is lost or no longer marketable. Solar tunnel dryers[11] are an inexpensive way of conserving such products without great losses or poorer quality (see Figure 2.5). On an area some 2m by 18m long and slightly raised off the ground, a foil can be stretched across like a pitched roof. Roughly half of the area serves solely as an air collector, with the other half containing the products to be dried by the air heated up in the collector. Small ventilators powered by solar energy can be used to ensure that the air circulates. This process not only reduces harvest losses, but also cuts the time required for drying roughly in half. Furthermore, the dried goods are protected from precipitation, which is common in tropical countries; the drying process continues without interruption during rain. And if the rain continues long enough to discontinue the supply of solar heat, a gas burner could provide auxiliary heat. Such a tunnel dryer can provide up to 60kWh of solar energy per day, equivalent to the energy contained in some 15kg of firewood.

Air collectors can also be used in temperate zones.[12] The energy yield is even higher than with flat-plate collectors for space heating. In energy-efficient buildings, air collectors and ventilation systems can be combined in a number of interesting ways.[13] For instance, at an indoor pool in Wiesbaden, Germany, a 420m^2 air collector heats the air inside, saving some 50,000m^3 of natural gas each year.[14] In Göttingen, Germany, a 343m^2 air collector on the wall of a heating plant pre-heats the combustion air with solar energy, reducing the amount of energy that the plant itself consumes.

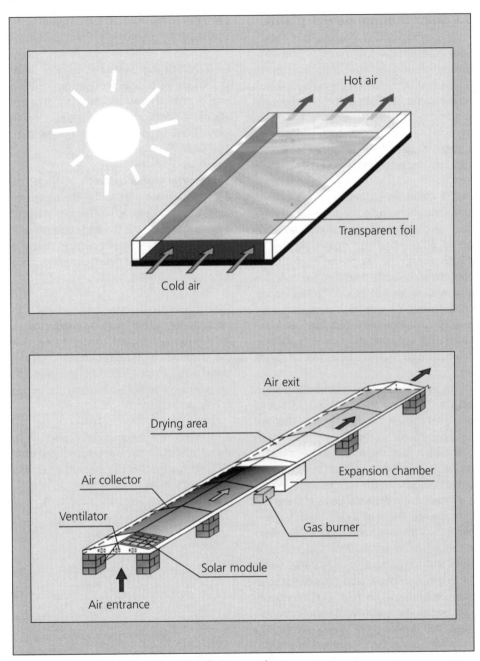

Figure 2.5 (1) Solar drying: How air collectors work
Figure 2.5 (2) Solar drying: Solar tunnel dryer for agricultural products

Source: W. Mühlbauer and A. Esper, 'Solare Trocknung', in Sonnenenergie 6/1997

2.6 Solar thermal power plants

Solar collectors can heat up to around 80°C. But much higher temperatures can be reached when sunlight is concentrated. This concentrated heat can then boil water and steam can be used to drive a turbine to generate electricity. In such cases, one speaks of solar thermal power plants or concentrated solar power (CSP).

In CSP plants, mirror systems track the sun and concentrate reflected light on a single focal point, where temperatures can reach up to 800°C. There, a fluid is heated to create steam. The steam can then drive a conventional turbine like the ones used in coal and oil power plants. Unlike photovoltaics (see Chapter 3), sunlight is not directly converted into electricity here, but rather indirectly via steam and a turbine. There are basically three types of CSP plants.

The most common type is **parabolic trough plants**. This technology has moved out of the pilot phase. In California's Mohave Desert, nine such power plants have been built with a total mirror surface area of 2.3 million square metres and a cumulative output of 354MW. The mirrors curve around a central tube at the focal point of the parabolic trough. The fluid inside these tubes is heated up to 400°C and used to generate electricity. Such commercial systems have been under construction since 2002, when Spain adopted special feed-in tariffs for solar thermal electricity. In mid-2006, construction began on a 50MW parabolic trough plant, and dozens more were planned.[15] In other Mediterranean countries and the US, large concentrating solar power plants are under construction or planned.[16] In the DESERTEC project, giant parabolic power plants are to be built in the Sahara to supply electricity to the north African countries and to Europe.[17]

In **dish Stirling systems**, a large dish-shaped mirror focuses sunlight on a Stirling engine, which uses the heat to generate electricity. With mirror diameters ranging from 7–17m, power output ranges from 10–50kW.

Solar tower plants concentrate sunlight on a single focal point like a dish Stirling system does, but they do so with a large number of mirrors that track the sun and concentrate it on the top of a tower, where steam is produced. Pilot systems currently have outputs of 1–10MW.[18]

CSP plants can generate up to 200MW of electricity per plant. A study conducted by the German Aerospace Centre found that the potential for Mediterranean countries alone was up to 1500TWh, equivalent to roughly half of all power currently consumed in Europe.[19]

And when the heat-carrying medium is stored, electricity can also be produced from solar energy even in the evening when the sun is not shining.

2

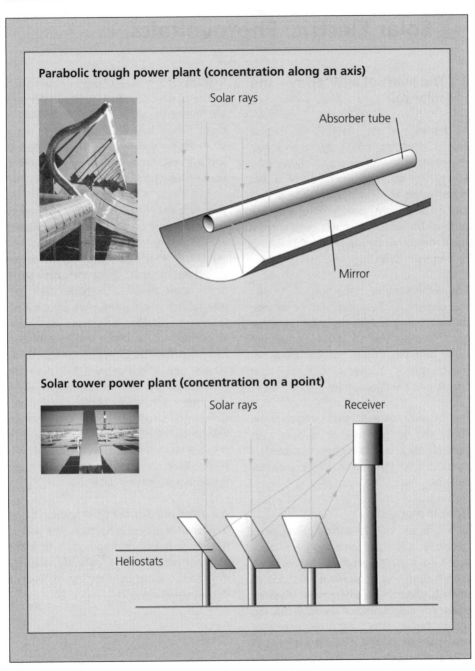

Parabolic trough power plant (concentration along an axis)

Solar rays

Absorber tube

Mirror

Solar tower power plant (concentration on a point)

Solar rays

Receiver

Heliostats

Figure 2.6 Solar thermal power plants

Source: The authors, Triolog

3 Solar Electric: Photovoltaics

3.1 The heart of a PV array – the solar cell

As described in Chapter 2, solar thermal energy converts solar rays into heat. Photovoltaics is another way of using solar energy. This technology is based on a well known effect in physics: some semiconductors convert light directly into electrical current. This chapter focuses on the main applications and the future of photovoltaics as seen from today.

Just as the absorber is the heart – the part that converts sunlight into heat – of a solar thermal array (see 2.1), the solar cell is the heart of a photovoltaic array. Spread out thinly over the largest possible area, this semiconductor converts incident light directly into electrical current.

The first solar cell was made using silicon in 1954, and 90 per cent of all solar cells currently made worldwide are still manufactured using this basic semiconductor material.

How it works

In a process called 'doping', impurities (generally boron and phosphor) are introduced to a wafer generally thinner than 0.2mm consisting of highly pure silicon to create two layers with different electric properties. When light reaches the solar cell, the charge carriers (electrons) from one layer flow to the other layer, creating a voltage of 0.5V at the contacts. This voltage within the solar cell remains relatively constant, but the current that comes out of the cell varies depending on the size of the cell and the intensity of incident light.

To reach voltage levels commonly used (such as 12V of direct current), multiple solar cells are 'switched in parallel' (connected in rows). When the cells are laminated into a sandwich between a glass pane on the top and a plastic foil on the back, you have a finished solar panel.

The efficiency of crystalline solar cells commonly sold on the market is generally around 15 per cent under standard test conditions. As temperatures rise, the power yield drops; ensuring that panels have some air underneath them to cool them off (natural ventilation) is therefore a good idea.

In addition to the silicon wafers described here, thin film has grown to cover roughly 10 per cent of the market. Thin-film cells generally consist of either amorphous silicon or some combination of copper, indium, gallium and selenium (CIS and CIGS), which will be discussed further in 3.5. Thin-film cells are much thinner than traditional crystalline silicon solar cells, and the production methods are also very different.

A number of other cell types are also being developed or in pilot production. The goal of this research and development is to make photovoltaics, which is currently relatively expensive, cheaper by using less material or less expensive material.[1]

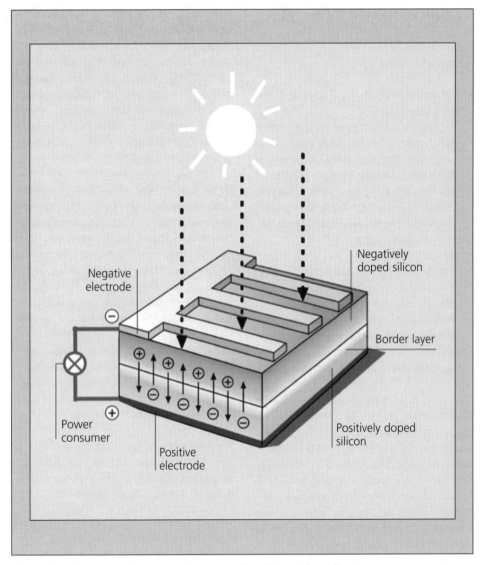

Figure 3.1 The heart of a photovoltaic array: The silicon solar cell

Source: Leuchtner and Preiser, *Photovoltaik-Marktübersicht*, 1994

3.2 Grid-connected PV arrays

Power generated from photovoltaics offers a number of benefits:

- Emissions – PV arrays are silent and emit no waste gases.
- Service life – since there are no moving parts, solar arrays have very long service lives. Manufacturers offer warranties of 20 years and longer for solar panels.
- Environmental impact – silicon solar cells are environmentally friendly during operation and can be recycled without any environmental impact.[2]
- Resources – silicon is the second most common element on the Earth's crust, so it is hard to imagine us running out of the raw material.
- Wide range of applications – photovoltaics can be used in a large number of applications from pocket calculators and wristwatches to large solar power plants.

Thanks to these benefits and a number of governmental policies to promote photovoltaics, grid-connected arrays have become a common sight on residential buildings in many countries (see Figure 3.2). Generally, they have a rated output ranging from 1–10 kilowatts-peak.[3] Solar panels produce direct current, which an inverter converts into alternating current so that the power can be exported to the grid. When solar power is metered separately from consumption (double metering), compensation for solar power can be decoupled from the retail rate (feed-in tariffs, see 11.10); this approach also allows solar power generation to be measured for carbon trading and towards renewables targets, which net-metering (with a single meter) does not.

A PV array with a peak output of around 2kW will take up some 20m² of roof space. Depending on local conditions and the array's orientation, that array will generate some 1700–2000kWh per year at German latitudes, roughly as much power as a four-person household that conserves energy would consume. In 2010, such an array would cost less than €8000, with prices falling rapidly (see 3.5).

When feed-in rates were offered for photovoltaics, businesses began to become interested. They began to set up much larger arrays with lower specific costs on multi-family dwellings, roofs of commercial facilities and farms. In 2006, arrays with a rated output ranging from 10 kilowatts-peak to 1 megawatt-peak made up around half of the German market.[4]

Today, the largest PV power plants have a peak capacity of dozens of megawatts. A 60MW array was completed in Olmedilla, Spain, in 2009 and there are plans to break the 100MW threshold in the US.

From Q4 2008 to Q3 2009, some 2.4GW of photovoltaics was installed in Germany, equivalent to the amount consumed by 720,000 German households.

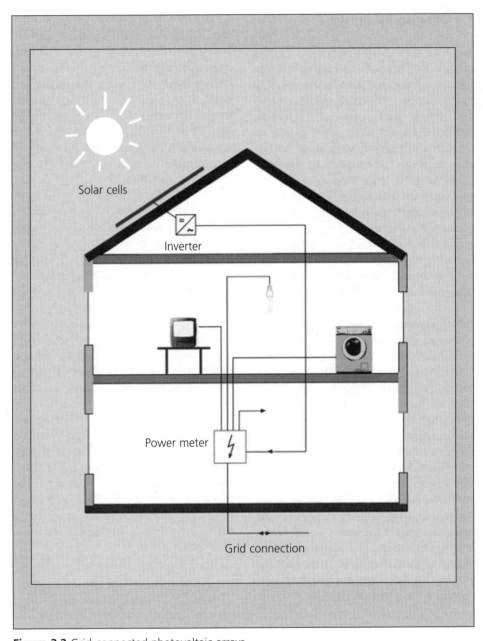

Solar cells

Inverter

Power meter

Grid connection

Figure 3.2 Grid-connected photovoltaic arrays

Source: The authors

3.3 Off-grid PV arrays

Roughly 2 billion people worldwide – a third of the world's population – have to make do without a power grid. Distributed electricity supply to cover the basics would improve the lives of these people considerably. Artificial lighting at night allows people on farms and in shops to work later; schools and community centres can also offer more flexible services. Radio, telephones and television provide information and means of communication to remote regions. The power needed for such applications can be provided less expensively and often faster with off-grid PV arrays than with grid expansion.[5] Indeed, photovoltaics is often even less expensive than small diesel generators.

The equipment used for such purposes is known as Solar Home Systems (see Figure 3.3), these consist of a solar panel roughly 0.5m^2 in size (around 50 watts-peak), a battery, a charge controller and three compact fluorescent light bulbs (12/24V). Under five hours of full sunlight per day, such as systems can power a radio, the three lights, a black-and-white television and other small devices for several hours. While that may not sound like much by European standards, it marks a crucial increase in the standard of living for many people in developing countries. Solar Home Systems for a basic power supply are available starting at around US$500 – a lot of money for the target group. Suppliers have therefore come up with suitable financing models.

For instance, Shell has the following strategy: 'You cannot expect a shepherd in Inner Mongolia to pay €500 or €1000 for a solar array. We have therefore come up with a kind of lease. We finance the system and install it. Customers then pay for monthly usage with a chip card that costs €7 for 30 days. They then have to recharge it, for instance at a Shell station or at their community centre. In this way, we get our investment costs back.'

The benefits of these PV systems are so obvious that this market is growing quickly without any subsidies. In 2005, more than 2 million households worldwide in developing countries received new Solar Home Systems. In China alone, some 10,000 new systems were installed per month in 2005.[6]

There are even situations in industrial countries where off-grid PV systems are the best option. For instance, PV-powered parking ticket meters are cheaper than the ones that run solely on batteries or require a grid connection. PV panels also make lighting possible for remote bus stops (with motion detectors to conserve electricity) to provide greater comfort and safety in public transport for rural areas. Farmers have also found that small PV arrays provide the cheapest power sometimes, such as to power a fishing pond aerator, an electrical fence and an irrigation pump.[7]

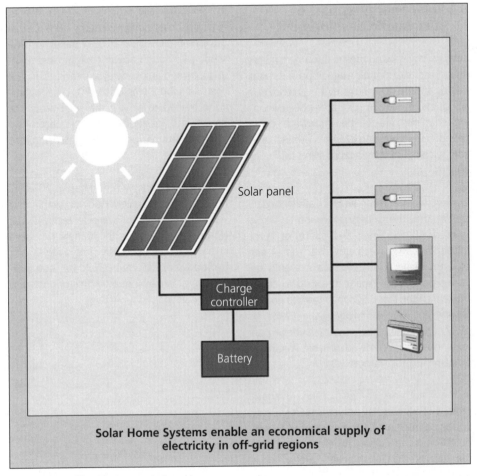

Figure 3.3 Off-grid photovoltaic arrays

Source: The authors

3.4 Solar energy as part of sustainable development

Solar energy is by no means the only requirement for sustainable development, but energy supply does have to be extended if standards of living in rural areas of developing countries are to be improved. Cuba provides an interesting example of how solar energy can raise standards of living for entire villages.

Comprehensive rural electrification began shortly after the Cuban revolution in 1959. From 1960 to 1992, the share of rural households with electricity rose from 4 per cent to 79 per cent. Because it would be very expensive to expand the grid to cover the remainder, some 500,000 remote homes still did not have electricity in the 1990s. In 1987, the Cuban government launched an ambitious programme to provide electricity to even the most remote villages.

In the first phase, all health clinics in these regions received solar power systems. Every village has such a health clinic staffed by one doctor and one nurse. This service is one of the reasons why Cuba's health care system is a model for Latin America. The solar array with an output of 400W powers a refrigerator for medicine and a small television. It also ensures that the clinic, which is often the main village meeting place, has sufficient light. By the summer of 2002, 320 such health clinics had received such solar power systems.

In a second phase, schools in remote villages received solar power systems. The Cuban government installed a 165W panel on 1900 schools – enough power to light classrooms and power a colour television and VCR. The cost per school was US$1480. Local people were trained to service the solar array, the battery and the devices. This solar electrification programme for schools was completed in 2002.[8]

In the third phase, which began in 2003, all homes were to receive power within five years. Most of them were to receive Solar Home Systems, with the panels produced locally in Cuba. In areas where wind power, hydropower or biomass could be used inexpensively, expansion into microgrids[9] was considered.

All schools in rural regions of Cuba use solar power

Figure 3.4 A Cuban village school with solar power

Source: BMU, Sustainable Energy Policy Concepts (SePco). Photo: Dorothee Reinmüller

3.5 The outlook for PV – lower costs from new technologies and mass production

Silicon-based photovoltaics is a well developed technology that reliably produces environmentally friendly solar power. At present, however, solar power from grid-connected arrays still costs around 30–40 eurocents in Germany per kilowatt-hour (depending on system size), roughly six to eight times as expensive as power from coal plants, though the price of solar power is a fraction of that in prime locations in places like Arizona. Research and development in photovoltaics has therefore mainly focused on lowering costs, mainly by means of two approaches that go hand-in-hand: new technologies and mass production.

New technologies will provide greater efficiencies and make do with less material or less expensive material. Various approaches are currently being pursued:[10]

Thinner cells

Over the past ten years, cells have become increasingly thinner – from 300 to 180 microns of expensive, highly pure silicon. But thin-film cells – mainly copper-indium-selenide (CIS) and cadmium telluride (CdTe) – make do with far thinner layers; CIS layers are only five microns thick, which represents a considerable reduction in material consumption. And because these layers are applied to a substrate by means of vapour deposition, manufacturing costs are also lower. Experts nonetheless believe that silicon cells will remain the main workhorse of photovoltaics.

Advanced designs

The surface structure can be optimized, for instance, to capture more light by reflecting less of it back outside the cell. The result would be greater efficiency.

Concentrator systems

If three types of semiconductor, each of which absorbs a different part of the light spectrum, are stacked on top of each other, efficiency increases dramatically. Such cells are, however, much more expensive. But if simple optical mirrors and lenses are used to concentrate light on such small cells, the energy yield increases even as material costs decrease.

Market development is also crucial in lowering costs. Industry will only be able to set up the large production capacity required for economies of scale if markets are sufficiently large to absorb output. The last decade clearly shows that greater solar panel production leads to lower panel prices (see Figure 3.5). Germany's 100,000 Roofs Programme, which ran from 1999–2003, and the more recent Renewable Energy Act (see 11.9) set off this development in Germany by providing a safe investment environment; the former with upfront funding, the latter with feed-in tariffs. Admittedly, the skyrocketing demand for solar panels led to a bottleneck in the supply of silicon in 2005, which temporarily kept module prices high.[11] But the silicon shortfall has been overcome, and prices for silicon solar panels fell by around 25 per cent in 2009 alone. Experts believe that grid parity – when solar power will cost the same as power from the grid – is already being reached in parts of southern Europe and the southwestern US, but even Germany is expected to reach grid parity by around 2013 because of its relatively high electricity prices.[12] German PV success is the direct result of its feed-in rates for solar power.

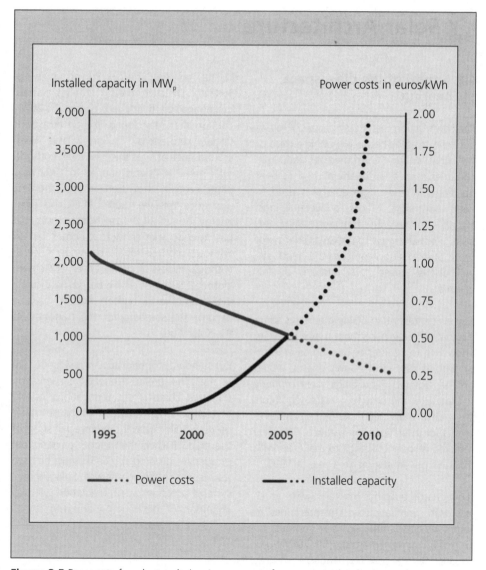

Figure 3.5 Prospects for photovoltaics: Lower costs from new technologies and mass production

Source: BMU, Erneuerbare Energien, 2006

4 Solar Architecture

4.1 A third of the pie – space heating

The construction sector plays a special role in our solar strategy because roughly a third of the final energy consumed in industrial countries such as Germany is used to heat buildings. Indeed, more energy is used to heat apartments, offices, public buildings, etc. than in the entire transport sector. Yet, there are many ways to use solar energy for these purposes (low-temperature heat). We will discuss these solar options in this chapter.

The average German building requires some 220kWh of heat per square metre per year, equivalent to 22 litres of heating oil or 22m^3 of natural gas. Most of this energy would not be required if buildings were properly insulated. In the past few years, legislators have fortunately begun setting up requirements for insulation. For instance, Germany required all new buildings to make do with 100kWh per square metre a year in 1995.[1]

'Low-energy buildings' make do with much less. Here, ventilation systems are added to improved insulation so that homes (in Germany) only need around 50–70kWh per square metre a year for heating – roughly 5–7 litres of oil, only a quarter of the average for buildings constructed before 1995. In Sweden, the low-energy standard was made mandatory for all new buildings at the beginning of the 1990s. Germany did not implement this standard until 2002.[2]

In its revised Energy Performance of Buildings Directive, the EU requires all new buildings to be 'nearly zero-energy' by 2020. Though that may sound like an ambitious target, zero-energy buildings have been around since at least the 1990s. The off-grid solar home (4.7) was opened in 1992 as a pilot project, but zero-energy homes, numerous passive houses (4.6) and plus-energy homes (4.8) constructed over the past two decades show that homes can do without conventional heating systems without any loss of comfort even at German latitudes. Such homes are becoming increasingly attractive for buyers; they will have to become the standard for the transition to the Solar Age.

But standards for new buildings will not take us far. The greatest energy conservation potential comes from renovations of old buildings. A solar strategy that focuses solely on new buildings will therefore fall short of the mark. Rather, the insights gained from progressive building standards must be used for renovation projects as well. Solar thermal systems and transparent insulation will play an important role here (4.4 and 4.5).

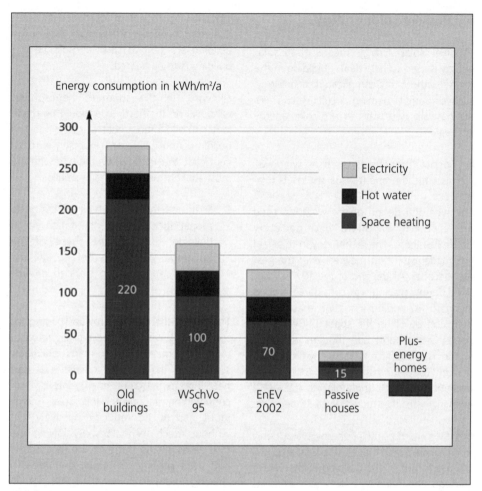

Figure 4.1 Space heating: A comparison of key energy figures in various building standards

Source: The authors

4.2 Passive solar energy

'Passive' solar energy means that solar energy is used to help heat a building in the winter without any complicated technology, such as pumps, heating circuits, etc. There are basically two ways to use solar energy passively:[3]

1   Special components, such as windows, glass façades and transparent insulation.

The type and design of the windows used are especially important because every house has them. The technology in this field has made significant progress over the past few decades. At the end of the 1970s, insulated double glazing was considered a good standard. By adding a coating that reflects heat and by filling the space between the panes with a noble gas, modern insulating windows can reduce heat loss by around 60 per cent (U_g = 1.2 instead of 2.8 for 'normal' insulating glazing[4]); triple glazing (U_g = 0.7) even increases that reduction to 75 per cent.

But if we are to use solar energy, we not only need to make sure that as little heat as possible passes out of the building in the winter through the windows; we also want as much solar energy as possible to enter the house through the windows. As the insulation of windows becomes better, the solar input is also reduced even as heat losses out of the building drop considerably. Nonetheless, south-facing, insulated windows have a positive energy payback. These windows allow more heat to enter the house than they allow to exit over the year.

Solar architecture takes advantage of this effect by reducing the building's energy demand with a greater share of south-facing glazing (in the northern hemisphere). North-facing windows cannot provide such heat input owing to a lack of direct sunlight. Therefore, windows on the north side of the building are kept as small as possible if a building is to be heated.

Likewise, in the southern hemisphere, windows on the north side would be made larger to allow more heat to come into the building. And wherever a building is to be kept cool, the windows on the side with the most sunlight would be made smaller.

2   Solar architecture, such as selecting the proper location and orientation for a building, adapting the shape of the building, using the roof to provide shading, and planting trees to provide shading.

The use of the roof to provide shading for south-facing windows is especially interesting (see Figure 4.2). In many buildings, sunlight in the summer creates excessive heat indoors, requiring energy-intensive air conditioning. To prevent this, solar energy input has to be reduced passively, i.e. without much technology. By having the roof eaves extend out over the façade, the roof itself provides a simple, but effective shading mechanism when the sun is high up in the sky during the summer. At the same time, the eaves should not extend out so far as to prevent direct sunlight from entering the building in the winter.

This example shows how traditional types of architecture use the sun. In the age of cheap oil, we have simply lost sight of a lot of these options. For the transition to the Solar Age, we need to fall back on this knowledge and combine it with new materials and technologies.[5]

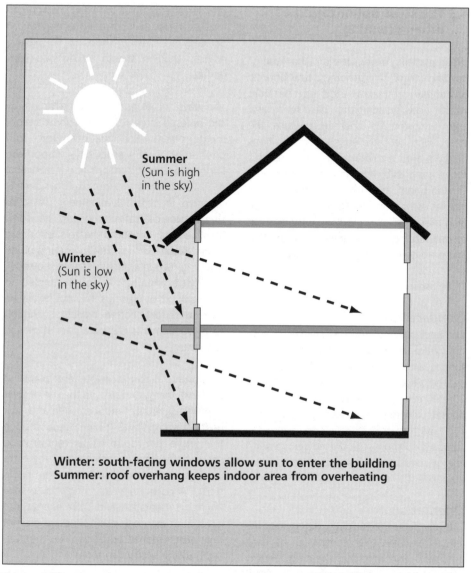

Summer
(Sun is high
in the sky)

Winter
(Sun is low
in the sky)

Winter: south-facing windows allow sun to enter the building
Summer: roof overhang keeps indoor area from overheating

Figure 4.2 Passive solar energy: South-facing windows and roof overhangs instead of heaters and air conditioners

Source: The authors

4.3 The solar optimization of urban planning

Urban planning experts decide what kind of building (single-family homes, row houses, apartment complexes, etc.) can be built where, how far apart they have to be and how they can be oriented towards the street. Until recently, energy considerations hardly played a role in such decisions. But today, it is possible to simulate future energy consumption and the potential of the passive use of solar energy during the planning process so that development plans can be optimized to reduce energy consumption and increase the use of solar energy.

Three aspects are especially important here:

Compactness
Compact buildings use less energy because they have a greater volume for a given surface area, where heat losses occur. For instance, five single-family homes consume 20 per cent more energy than five row houses with the same floor space. In turn, five apartments of the same size in a compact complex would lower heating energy consumption by as much as a further 20 per cent.

Orientation
The orientation of buildings determines the extent to which solar energy can be used passively. For instance, heating energy demand increases by as much as 15 per cent if a low-energy house is poorly oriented. And an optimal southern façade is even more important for passive houses (see 4.6).

Shading
Most heating energy is needed in the cold season, when the sun is the lowest in the sky – and the buildings tend to shade each other the most, thereby preventing the passive use

of solar energy. Over the year, shading can increase demand for heating energy by 10 per cent in low-energy housing. For passive houses, shading should be avoided altogether.

The ways and extent to which solar energy can be used largely depend upon development plans. Specifications for roof orientation are especially important. Furthermore, specifications about individual or district heating systems are often already specified in urban development plans, so that ecological district heating networks (cogeneration, woodchip heaters, etc.) have to be taken into consideration during urban planning. Urban planning cannot, however, specify everything – for example, in Germany urban planners may not legally be able to require better insulation, though such requirements can be a part of private contracts.[6]

Optimizing urban development plans to conserve energy and step up the use of solar energy generally reduces demand for heating energy by 5–15 per cent, though that figure may rise to 40 per cent in some cases. In the light of such savings, the extra costs for more careful planning seem negligible.[7] A community that neglects to take energy conservation and solar energy into account in its urban planning therefore not only wastes money, but also makes the transition to the Solar Age harder.

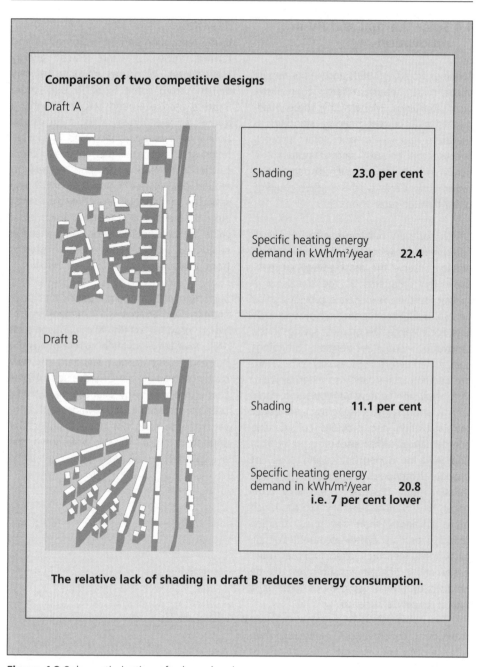

Comparison of two competitive designs

Draft A

| Shading | **23.0 per cent** |

| Specific heating energy demand in kWh/m²/year | **22.4** |

Draft B

| Shading | **11.1 per cent** |

| Specific heating energy demand in kWh/m²/year | **20.8** |
| | **i.e. 7 per cent lower** |

The relative lack of shading in draft B reduces energy consumption.

Figure 4.3 Solar optimization of urban planning

Source: Deutsches Ingenierblatt, Sept 1996

4.4 Solar thermal and PV in renovation

Buildings from the 1960s and earlier require much more heating energy than more recent buildings. Modernizing these buildings to fulfil current building standards is therefore crucial in our solar strategy because the greatest savings potential is here – in Germany, roughly 20 per cent of overall current energy consumption could be offset through renovation.

Of course, every building has to be investigated individually to determine which steps would conserve the most energy inexpensively. But regardless of the details of a specific building, experience provides some general guidelines for such renovation projects. For instance, the greatest savings in the renovation of residential buildings constructed before 1980 usually comes from the addition of insulation to external walls. The installation of new windows and doors with a special insulating glass and better roof insulation also provides considerable energy savings. While insulating the bottom floor and the uppermost ceiling does not provide tremendous energy savings, these measures are, however, often very profitable. New heaters generally also run much more efficiently than old ones, thereby reducing primary energy demand. Overall, all of these measures taken together often easily allow energy consumption to be reduced by 50–70 per cent or more in a typical residential building.

Additional conservation measures then often become very expensive, but the use of solar energy can then be a good option. A solar thermal array can help reduce energy consumption even further.

Example

In the renovation and expansion of an end terrace house from the 1920s,[8] energy conservation measures reduced energy demand from some 320kWh per square metre a year to around 80kWh, roughly a quarter of the original value even as the floor space was greatly increased. The installation of a large solar thermal array (23m^2) further cut heating demand in half to around 40kWh (see Figure 4.4). While the savings from the solar array only amount to around 12 per cent in terms of the original value for heating demand, comprehensive renovation increases the share of savings from solar energy.

Such renovations require upfront financing, and the German KfW Bank, a state organization, provides low-interest loans for such projects. If the renovation project brings the building to the current standard for new buildings, the bank underwrites up to 15 per cent of the loan.[9] Any solar thermal systems used are included in the savings, so if renovation alone does not bring a project far enough to qualify for the write-down, the energy offset by solar energy is counted.[10]

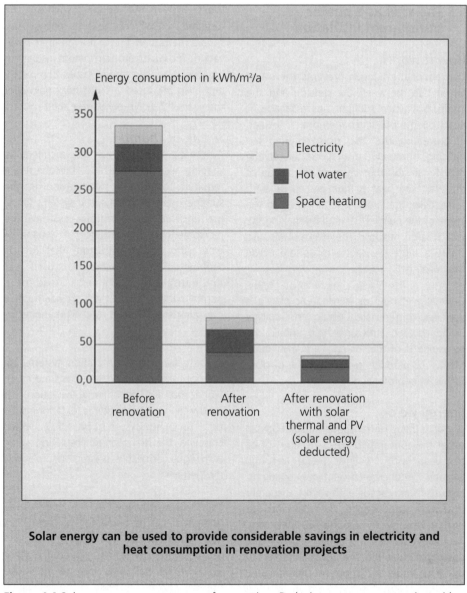

Solar energy can be used to provide considerable savings in electricity and heat consumption in renovation projects

Figure 4.4 Solar arrays as a component of renovation: Reducing energy consumption with and without solar arrays

Source: Ranft and Haas-Arndt, Energieeffiziente Altbauten, 2004

4.5 The wall as a heater – transparent insulation

How it works

Conventional insulation prevents the transfer of heat between the outside and the inside. But when sunlight hits a façade, it heats up the building's membrane. Instead of preventing this heat from entering the building, transparent insulation channels the heat to an absorber layer on the inside of the wall. The heat is then passed on with some delay (generally several hours) to the inside of the building, essentially making the wall a large radiator. The warm walls also create a more pleasant atmosphere inside the building.[11]

Generally, aerogel or transparent plastic is used in such transparent insulation. Capillary or honeycomb structures help enable the combination of transparency and insulation. Most such products require a pane of glass for weather protection.

Energy yield

A square metre of transparent insulation can offset the consumption of 5–40 litres of oil per year, with 15 litres being a rough average. On energy-efficient new buildings, a south-facing façade with such transparent insulation can reduce the demand for heating energy by around 20 per cent. Renovated buildings also benefit from the additional insulation. A renovation project at the Paul Robeson School, a prefabricated concrete structure in the German town of Leipzig, provided some interesting results. Here, 300m^2 of transparent insulation was installed, and although the sun is not very intense in the winter, heat was still provided in those months.

Walls with conventional insulation still required 43kWh of heating energy per square metre over the year even after renovation. The walls with transparent insulation, however, reduced these losses completely and even produced a slight heat gain over the year of 2.3kWh per square metre.[12]

A rule of thumb

Automatic shading systems or architectural shading elements (such as balconies) are required to prevent overheating in the summer. Because mechanical systems (lines) malfunction easily, simpler systems are recommended. For instance, if a pane of glass with a prism structure that reflects summer sunlight in particular is used over the transparent insulation instead of a normal pane of glass, it will provide roughly the same shading as a roof overhanging by 1m.[13]

Because transparent insulation systems are still not mass-produced, the costs are much higher than for conventional insulation, but sandwich insulation systems (exterior insulation finishing systems or EIFS), which combine the transparent insulation plate with light-permeable plaster, are relatively inexpensive.[14]

Transparent insulation has been under development since the beginning of the 1980s, and hundreds of buildings have such façades in Germany covering some 50,000m^2 of wall area. Transparent insulation will play a major role in the renovation of buildings in the future.[15] One important new application is the use of transparent insulation to detract light in daylighting elements used in gymnasiums, museums, industrial halls, etc.

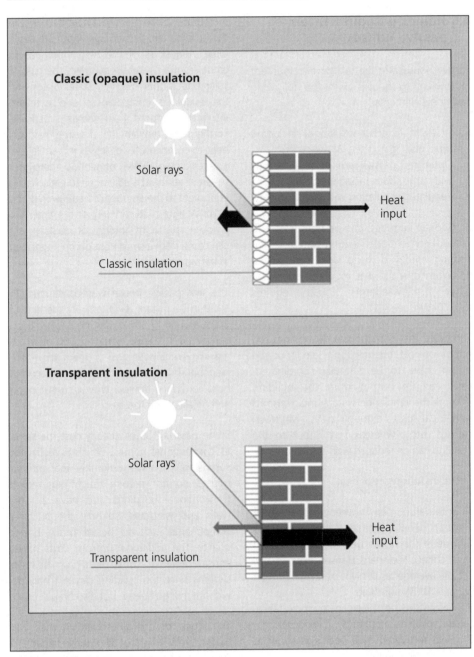

Figure 4.5 Walls as heaters: Transparent insulation

Source: BINE: Transparente Wärmedämmung zur Gebäudeheizung, Bonn 1996

4.6 Homes without heaters – passive houses

Passive houses are the further development of low-energy houses with the following additional elements:

- Excellent external insulation, including triple glazing.
- Optimized passive use of solar energy with large south-facing windows.
- Regulated ventilation with heat recovery.

At German latitudes, passive houses can do without conventional heating systems. Their heating energy demand does not exceed 15kWhm² over the year, roughly a quarter of what a low-energy house requires (50–70kWh, see 4.1).

To ensure that such homes do not get too hot or too cold, the building's heat input and output have to be carefully coordinated. Heat can be output from the building through the ventilation system and transmission (heat losses through walls, windows, ceilings, etc.), whereas heat input to the building can be reduced with insulation.

There are four types of heat:

1　Internal sources, in other words the heat that people, animals and appliances within the house generate. As German architect Meinhard Hansen likes to say, the average adult gives off as much heat as a 100W lightbulb.
2　The passive use of solar energy, for instance, the solar heat input through windows.
3　Heat recovered by a heat exchanger in the ventilation system.
4　An auxiliary heater when the other heat sources do not suffice.

In old buildings, especially those from the 1960s, heat losses from the building's membrane are so great that heat from within the building and passive solar input are negligible. Almost all of the heat input has to come from a heater. In contrast, the passive solar input is optimized in passive houses, and the building's compactness and excellent insulation reduces heat losses drastically. Furthermore, the ventilation system recovers heat to compensate for losses in ventilation. The remaining 15kWh of heating energy per square metre that is still required each year is equivalent to the amount of heat given off by a 100W light bulb running all the time.[16] In practice, this slight amount of residual heat can come from a small heat pump integrated in the ventilation system.

The first passive house was constructed in 1999 in Darmstadt, Germany. By the end of 2009, around 17,500 passive houses had been completed in Europe, some 13,000 of which are in Germany. Several of them are entire neighbourhoods. Scientific follow-up studies have confirmed the low energy consumption and excellent comfort standards.[17]

Some passive houses already cost the same as conventional homes, which is really not surprising when we remember that passive houses do not require much high-tech.[18] Conventional buildings also have a roof, walls and windows, so only the building components have to be of much better quality. The additional upfront costs offset considerable upfront costs for heating systems since the building can not only do without a large heater, but also fireplaces.

The share of energy-efficient homes will quickly grow in light of all of these benefits.[19] Different types of passive houses are going up all over Europe – as stand-alone single-family homes, row houses or apartment complexes. Indeed, nurseries, schools, gymnasiums and office buildings have also been constructed as passive houses.[20]

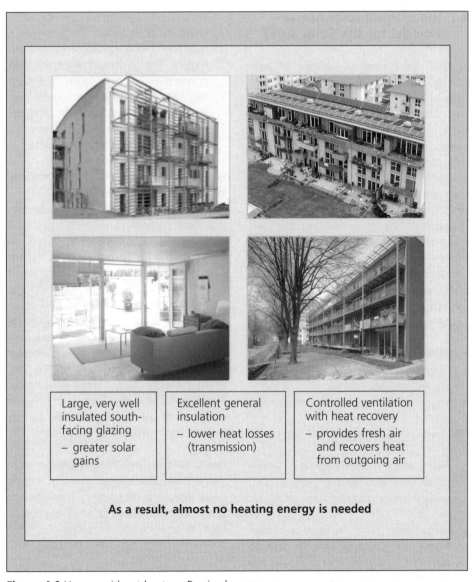

Large, very well insulated south-facing glazing
– greater solar gains

Excellent general insulation
– lower heat losses (transmission)

Controlled ventilation with heat recovery
– provides fresh air and recovers heat from outgoing air

As a result, almost no heating energy is needed

Figure 4.6 Homes without heaters: Passive houses

Source: Phasea, Fraunhöfer Institute for Solar Energy Systems, Ralf Killian

4.7 The off-grid solar house – a model for the Solar Age?

In 1992, the Fraunhöfer Institute for Solar Energy Systems opened its Energy-Autonomous Solar House in Freiburg, Germany, as a research and demonstration project.[21] With 145m² of floor space on two stories, it has five rooms, a kitchen and ancillary rooms. This house proved that a single-family home could make do with the solar energy from its roof and walls the whole year even in the German climate. The 'autonomy' here thus concerns not only space heating, but also hot water, gas for cooking, and electricity. Several years of operation demonstrated that this independence is possible without any loss of comfort for residents.[22] Like a passive house, this house has:

- Extremely good insulation.
- Large southern windows to exploit solar energy passively.
- Regulated ventilation with heat recovery.

The following technologies were also used:

- Highly efficient solar thermal collectors (14m² with a 1000 litre storage tank) provide enough hot water almost all year round.
- Large areas of transparent insulation (some 70m²) reduce heating demand down to 0.5kWhm² over a year – only around 1 per cent of the energy needed in a low-energy house and around 4 per cent of what a passive house needs.
- A photovoltaic array (4.2 kilowatts-peak) generates some 3200kWh of electricity per year, more than the efficient appliances used in the household (60 per cent less energy consumption than with conventional appliances) need. Excess electricity is used to convert water into oxygen and hydrogen, with the latter being stored in tanks. The hydrogen is used for cooking and auxiliary heat a few days a year. When the photovoltaic panels have not been receiving enough sunlight, the hydrogen can also be used to power a fuel cell, which generates electricity. The waste heat created in the process is used to heat service water.

Is the off-grid solar house a model for the Solar Age?

The Energy-Autonomous Solar House does not have a grid connection, nor is it connected to a district heating network. As such, crucial solar options – district heating with solar energy or biomass and the use of other renewable energy sources, for instance, are not possible here. As a result, this project required a large amount of seasonal storage that would not have been necessary if the house had been connected to public heat and power sources. The Solar House therefore did not show what houses would look like in the Solar Age, rather, it showed that a house in Germany can do without fossil fuel even in the 1990s

Finally, the Energy-Autonomous Solar House also showed that solar architecture is not a single technology, but rather a cornucopia of technologies that have to be closely coordinated to produce optimal results.

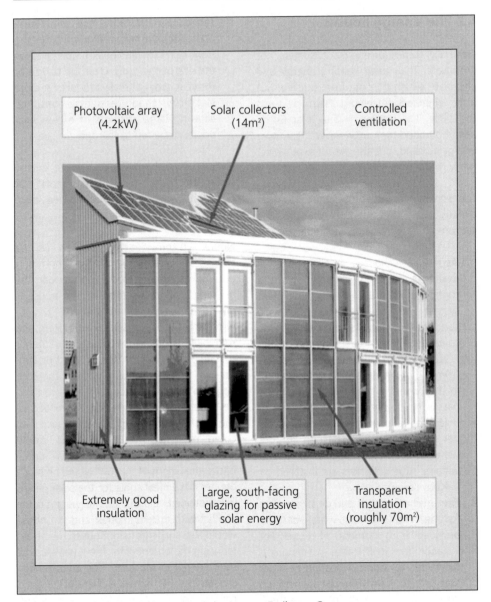

Figure 4.7 The Energy-Autonomous Solar House, Freiburg, Germany

Source: Fraunhöfer Institute for Solar Energy Systems

4.8 Plus-energy houses

Generally, homes are energy consumers, not producers. They need heating energy and electricity, which they get from outside. But a model project in Freiburg, Germany, turns the table. The houses in the Solar Estate neighbourhood not only cover their own energy demand, but also export excess electricity to the grid.[23]

The technologies in this residential area have already been presented in the previous sections. They were simply recombined here.

- As in passive houses, excellent external insulation on the walls, windows and roofs is combined with a ventilation system with heat recovery and large south-facing windows that passively use solar energy to cover the remaining residual energy demand, which ranges from 6–12kWh per square metre a year in this project – considerably below the passive house standard (15kWh).
- Solar thermal collectors provide up to 60 per cent of the heat needed for hot water; to keep the roofs clear for photovoltaics, tube collectors are installed on balcony railings.[24]
- The small amount of energy still needed for heating and hot water can be provided in a number of ways. For instance, the Solar Estate in Freiburg is connected to a district heating network that gets its heat from a cogeneration unit. A similar project in the southern German town of Ulm was to use a stove fired with wood pellets in an automatic feed system (for more on wood pellet stoves, see 5.4), though the project never came to fruition.

- The large roofs facing the south have photovoltaic arrays with outputs ranging from 4–7 kilowatts-peak (depending on the size of the roof). Over the course of the year, they generally produce far more electricity than residents consume, allowing large amounts of excess solar power to be fed to the grid.

Overall, these homes export more energy (solar electricity) than they import (for heating). They are therefore called plus-energy homes.

Figure 4.8 provides an overview of the various components used in these homes. It also makes another crucial aspect clear: the asymmetrical cross-section of the house is no accident, but rather a useful way of improving the use of solar power. Here, the roof peak is not in the middle, but rather towards the northern side of the building, making the south-facing part of the roof larger so that the photovoltaic array can also be bigger. If the roof peak were moved all the way back to the north, the space that could be used for solar panels would increase, but there would also be two drawbacks: the height of the roof would be much greater if the ideal angle for the solar panels were retained, and that greater height could conflict with building codes; and the house would also shade its neighbour to the north, reducing the potential for the passive use of solar energy in the back. The cross-section chosen in these homes is therefore a good compromise between a large PV area and low shading of the neighbours.

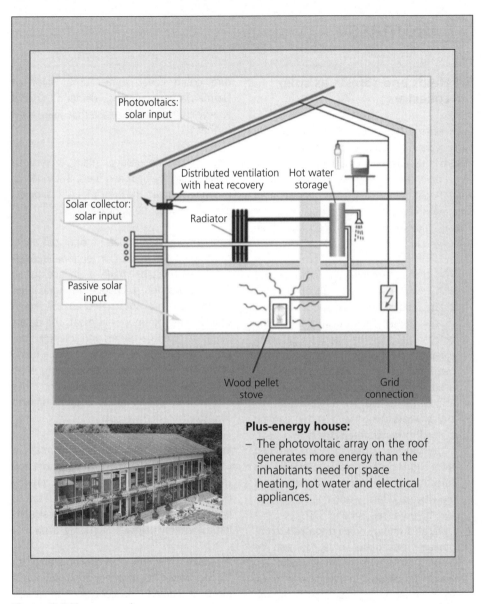

Plus-energy house:
- The photovoltaic array on the roof generates more energy than the inhabitants need for space heating, hot water and electrical appliances.

Figure 4.8 Plus-energy homes

Source: Based on Disch. *Photo:* Michael Eckmann

5 Biomass

5.1 Fields and forests as solar collectors

Solar energy has always been used on a large scale in agriculture and forestry. Plants absorb the energy in sunlight and store some of it chemically – as biomass. When animals eat these plants, another type of biomass is created as a byproduct: manure. Overall, biomass can be divided into three categories in terms of energy use:

1   Wet biomass (manure in particular, but also freshly cut plants) can be used to create biogas when allowed to ferment in an oxygen-free environment. The biogas can then be used to generate electricity, heat or both. Waste gas from trash and water purification can also be used in this way, and organic waste can also be fermented (see 5.2 and 5.3).
2   Dry biomass (wood and straw) can be burned to generate electricity and heat. In addition to straw (a waste product from agriculture), waste wood from forestry and timber byproducts from industry (sawdust, wood chips, etc.) can also be used (see 5.4 and 5.5).
3   Dedicated energy crops (rapeseed, corn, poplars, miscanthus, etc.) can also be planted to provide additional biomass, which can be used in a number of ways to make electricity, heat and fuels (see 5.6–5.9).

Wet and dry biomass are generally a byproduct or waste product of current industrial processes; therefore they generally need to be disposed of as waste. Using them as a source of energy would allow this material to be recycled, thereby reducing waste. Waste wood, straw and biogas from agriculture could cover some 5 per cent of Germany's primary energy demand.[1] Special energy crops would increase that share even further.[2]

Biomass can be used in many ways – to make electricity, heat and fuels. But while its potential is great, the principles of ecological agriculture must be upheld.

Furthermore, a number of principal limits must be kept in mind. Conventional agriculture itself must be made more ecological. The reasons for a change in agriculture include, but are not limited to, nitrate in groundwater, hormones in meat, the disappearance of entire species and soil pollution. Extensive agriculture, which requires a much larger area for the same production, is an important step in the right direction,[3] but this requirement limits the amount of land available for energy crops.

In addition, plantations and harvests of energy crops require energy for tractors and the irrigation. They first consume energy before producing it. This energy input has to be taken into consideration when calculating the energy payback of energy crops.

A number of energy crop concepts only pay for themselves because they are highly subsidized. If agrarian policy was restructured with a focus on ecology, the number of energy crops available would be reduced.

5

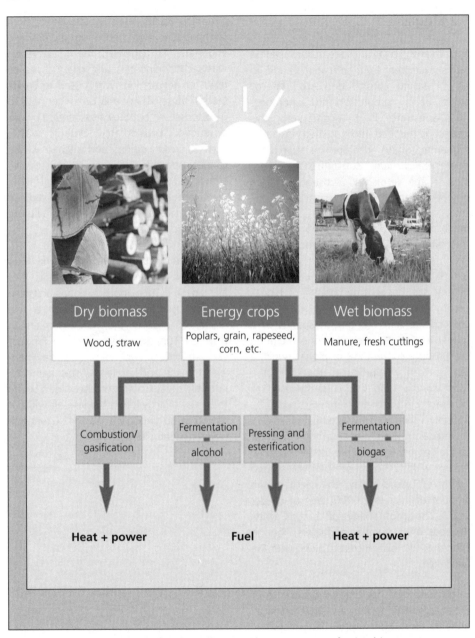

Figure 5.1 Forests and fields as solar collectors: The many ways of using biomass

Source: The authors

5.2 Biogas

In Germany, biogas units are mainly used on farms. Excrement from farm animals and, to an increasing extent, dedicated energy crops, are first ground up into a homogenous substrate. In a second step, this mixture is pumped into a heated, insulated fermenter, where anaerobic bacteria break down the organic substance to create biogas. The fermented substrate is then pumped into a storage tank.

Biogas consists of around 65 per cent methane and 30 per cent carbon dioxide (CO_2). Its energy content is around 6–6.5kWhm³, compared to 9.8kWh for natural gas. The energy content of biogas greatly depends upon the substrate's composition and the time it spends in a fermenter. For instance, a ton of cattle dung will produce some 45m³ of biogas, while a ton of corn will produce around 180m³. Most of this biogas is currently used to generate electricity in Germany (see 5.3). Roughly a third of the waste heat created in the process is used to heat up the fermenter to the optimal temperature for the bacteria. A farm with 120 cows can produce around 100m³ of biogas a day, equivalent to the amount of heat in 23,000 litres of oil per year.[4] The profitability of biogas units depends on a number of factors, though economic feasibility increases along with the unit's size.

Biogas units are not only used because of the energy benefits; they also provide special benefits for agriculture. For instance, fermentation improves the value of the manure used as fertilizer by making it more palatable to plants. In addition, the substrate is more homogenous and can therefore be pumped more easily. Finally, manure does not have such a strong smell after fermentation when sprayed onto fields.

Organic waste and wastewater from kitchens are also becoming increasingly important. In particular, the focus is on waste containing fats and oil, such as oil used for deep-frying. When used to create biogas, these oils not only no longer have to be disposed of, but they also increase biogas production considerably. Organic waste, such as fresh cuttings and kitchen waste, cannot, however, simply be combusted like straw and wood because of the higher water content. But when they are mixed into a biogas slurry, they can be used without further ado.

At the end of 2008, there were some 4000 biogas units in Germany,[5] and roughly twice as many in Europe. Indeed, biogas units can even be used in cities. For instance, a new residential area in the northern German town of Lübeck widely uses the kind of vacuum toilets commonly seen in high-speed trains and airplanes. This approach not only saves water, but also allows waste to be fed directly into a biogas unit without having to go through a sewage system first (www.flintenbreite.de).

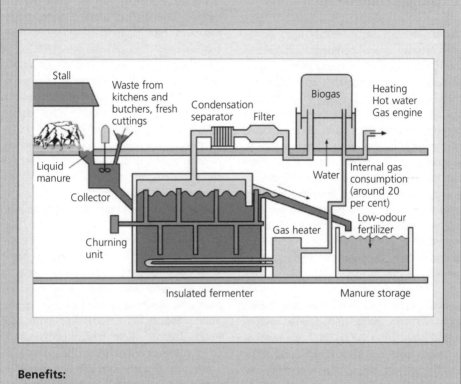

Stall

Waste from kitchens and butchers, fresh cuttings

Condensation separator   Filter

Biogas

Heating
Hot water
Gas engine

Liquid manure

Collector

Water

Internal gas consumption (around 20 per cent)

Churning unit

Gas heater

Low-odour fertilizer

Insulated fermenter

Manure storage

Benefits:
- **Biological waste is used as a source of energy**
- **Agricultural waste can be better used as fertilizer**
- **Odours are reduced**
- **New revenue stream for farmers**

Figure 5.2 Diagram of a biogas unit

Source: Energie für helle Köpfe

5.3 Biogas cogeneration units

Under German law, electricity from biogas units receives special compensation (feed-in tariff). If a biogas unit is then run as a cogeneration unit, the generator's waste heat can then not only be used to heat the fermenter, but also to provide space heat. The power generated can be used internally or exported to the grid. In the latter case, around 10 eurocents per kilowatt-hour is paid under current German law (see 11.10). If dedicated energy crops are mixed in with liquid and solid manure to create biogas, a 6 per cent biomass bonus is also paid. The financial incentive is so attractive that almost all biogas units in Germany are used mainly to generate electricity.

Germany's Renewable Energy Act, which took effect in 2000, gave biogas units a strong push. From 2000–2009, the number of biogas units rose fivefold to around 4800, and the average output of new systems rose from 75kW to 350kW from 2000–2003.[6] The biomass bonus stepped things up even further.[7] The main energy plant used is corn. For each kilowatt of installed capacity, roughly 0.5ha (around 1.25 acres) of corn needs to be planted[8] (see 5.6 and 5.10).

Cogeneration units that use biogas are also remarkable for two reasons. First, their carbon balance is excellent. Biogas is itself carbon-neutral because only the amount of carbon that the plants took out of the atmosphere is emitted when the substrate is fermented and the gas combusted. Furthermore, when this gas is used to power a cogeneration unit, electricity from another source (such as a coal plant) is offset, leading to carbon reductions there. Finally, whereas methane, a heat-trapping gas, enters the atmosphere when biomass is allowed to decompose, it does not when used to generate electricity. In the process, the biogas cogeneration unit's climate impact improves further. Overall, emissions from biowaste are reduced by some 100kg of greenhouse gases when used in a cogeneration unit rather than used to make compost.[9] A biogas cogeneration unit is therefore not only carbon-neutral during operation, but also actually reduces greenhouse gas emissions in the larger picture, making these units carbon and methane sinks.

The second thing that makes cogeneration units remarkable is the special role they play in a solar energy supply. As is well known, electricity is hard to store. Concepts for a solar power supply therefore have a problem: solar and wind power naturally fluctuate, requiring additional power generation capacity on cloudy days with little wind. This is where biogas cogeneration units come in. After all, biogas is easy to store, so that such units can be ramped up and down as the grid demands. In other words, biogas can compensate for some of the fluctuations in solar and wind power.

Over the long term, additional storage and control techniques will be needed to adjust the supply of solar energy to demand. Biogas cogeneration units will play a role here (see 10.1).

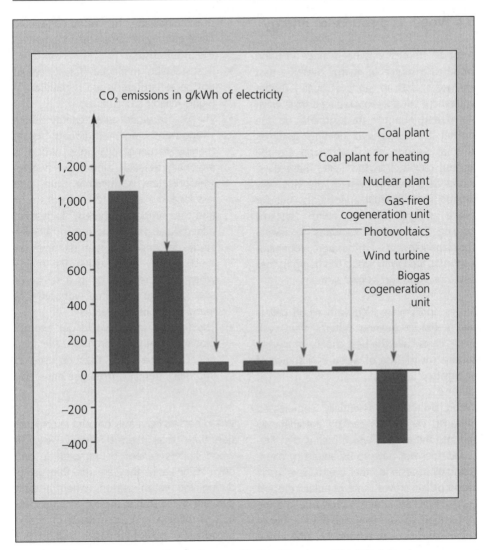

Figure 5.3 Biogas cogeneration unit lowers CO_2 by offsetting other CO_2-intensive generation sources

Source: Authors' representation based on GEMIS, Öko-Institut e.V.

5.4 Wood as a source of energy

A lot of progress has been made in using wood as a source of energy over the past few years, both in terms of quality (better equipment considerably reduced toxic emissions) and quantity (fast growth on the market for wood-fired heating systems). Austria, Sweden and Switzerland are the leading countries in this field. These days, wood-fired heating systems are not only offered for individual apartments, but for entire neighbourhoods, with systems ranging from several kilowatts to several thousand kilowatts. Furthermore, cogeneration units fired with wood gas being further developed and run as pilot systems.

Stoves and boilers fired with wood pellets and a district heating systems fired with wood chips have the best chance of market success for the use of wood as a source of energy (see 5.5).

Wood pellets are essentially compressed sawdust. The pellets can be automatically fed into the burner, which means that firewood does not have to be added by hand. Rising oil prices and other events have made wood pellets stoves quite popular in recent years. Pellets are currently produced in some 30 locations in Germany, and the number of systems has risen to around 125,000 in the year 2010.[10] Pilot projects have demonstrated that wood pellets can also be used to generate electricity and heat in cogeneration units.[11]

Wood energy offers a number of crucial benefits:

- Wood is a carbon-neutral fuel. If the amount of wood we take out of the forest never exceeds the amount that can grow back (in line with the principle of sustainability), then wood fuel would never emit more carbon than the trees in the forest absorb.
- If sustainably managed, forests would not be exhausted. Wood is therefore a highly reliable source of energy.
- The use of wood as an energy source strengthens regional added value, thereby securing local jobs. Wood is regionally available almost everywhere. Transportation is therefore short, and less local money has to be spent on oil and gas imports, thereby increasing added value within the region. A Swiss study demonstrated that six times as much money stays within the region when wood replaces oil as a source of energy. At the same time, eight times as many local jobs are created.[12]
- The forestry sector would get another global market for its products. The additional revenue could then be used to improve forest management, for instance.

Wood heaters can easily be used in conjunction with solar thermal energy. Because wood boilers cannot be ramped up and down very easily, they are not that useful during the warm season, when heating demand is low. During these months, a fossil fuel is generally used instead of wood because such boilers are easier to adjust. But a solar thermal array can complement a wood heater very well. The former provides heat during the summer; the latter, during the winter. Over the year, you then get all of your heat from renewables.

In Austria, and increasingly in Germany, such systems are becoming popular. Wood heaters in combination with solar arrays can also be used in connection with district heating networks.[13]

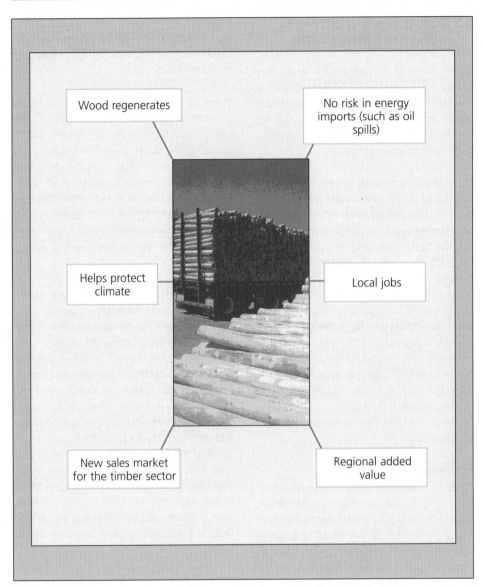

Figure 5.4 Getting energy from wood

Source: The authors

5.5 District heating networks with woodchip systems

District heating networks with wood-fired boilers offer tremendous opportunities for the fast expansion of wood in the heating sector. In this scenario, wood covers the baseload, with a gas or oil boiler covering peak demand on cold days.[14]

For example, a heating centre with both a wood and an oil boiler were installed in a new residential area with 100 single-family homes. The wood boiler was dimensioned to cover heating demand on a normal winter day. When extreme cold fronts come through, the oil boiler is switched on to cover peak demand. The wood is delivered as chips some 5–10cm long. These chips are automatically fed into the boiler by means of a screw conveyor. A 50m³ silo can contain enough fuel for around a week. The district heating network has well insulated lines to transport heat from the heating centre to homes, where it is passed on to heating systems and hot water tanks via heat exchangers.

Emissions

When we compare heating systems, emissions that occurred during fuel extraction, processing, and transport have to be taken into consideration. If we consider a wood-supported district heating network (with an oil/gas oil boiler for peak demand), we get the holistic values shown in the chart to the right.[15]

Because the impact of various types of pollution is not always comparable, the limit values in German law (TA-Luft) were weighted and then added up in line with the principle that a pollutant has a lower limit value the more toxic it is. This approach allows us to approximate a comparison of emissions from various heating systems (see Figure 5.5). Here, wood heating systems perform worse than oil and gas heaters in terms of pollution.

However, would heaters perform far better in terms of carbon emissions. Indeed, the carbon emissions from the system with a wood heater mainly come from the oil-fired boiler that covers peak demand. A look at the absolute amounts shows how much better wood performs. While the gas-fired heating system emitted 2.5 fewer tons of pollution, it also produced 202 tons more CO_2 than the wood-fired heating system did!

Recent discussions about particulate matter have also brought wood stoves back into the spotlight. There can be no doubt that wood-fired heating systems give off smoke – and hence particulate matter. However, the data used in such debates often pertain to open fireplaces and outdated equipment, not modern, low-emission wood stoves. These modern systems actually have lower particulate emissions than normal oil heaters when running full blast with standard wood as a fuel. And while emissions do increase when the system is running at partial capacity, a buffer tank can be used to reduce emissions.[16] Finally, electric filters can be used to reduce particulate matter from woodchip systems.

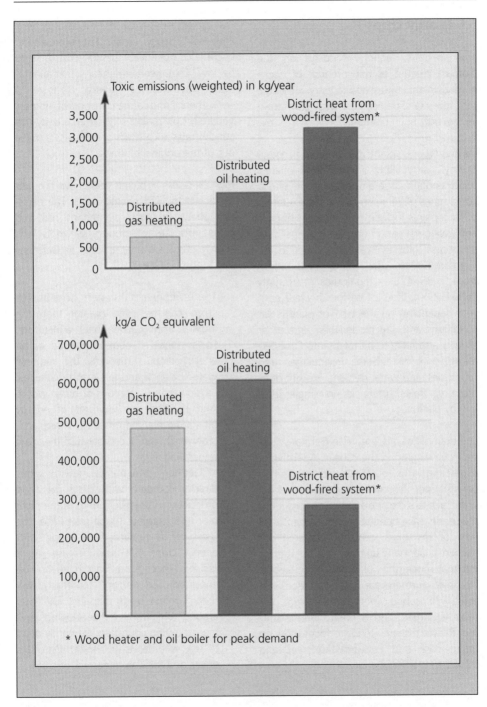

Figure 5.5 District heating systems with woodchip heaters: A comparison on emissions

Source: Forstabsatzfonds, Holzenergie für Kommunen, Bonn 2003

5.6 Energy crops

The previous sections focused on the biomass created as a byproduct or waste product in agriculture and forestry. But in the past few years, dedicated energy crops have increasingly been planted.

Various types of plants can be used to 'grow energy', particularly lignocellulosic plants, which contain large amounts of energy-rich compounds of lignin and cellulose. Examples of such plants include trees (such as poplars and willows) and grasses (grains and subtropical grasses like miscanthus). In the German climate, the energy yield ranges from 85–425 gigajoules (roughly 2400–12,000 litres of oil) per hectare each year depending on the type of plant, solar conditions and specific location. But other aspects also have to be taken into consideration, such as labour, insecticides and fungicides, and water demand. In light of all these variables, there is no single ideal energy plant.[17]

The example of miscanthus in Germany illustrates how complex the issue is. A relative of sugarcane, this plant can produce rich harvests per hectare, which led to high expectations a decade ago in Germany.[18] But the results have been sobering. Some 20–25 tons of dry mass can be harvested per hectare in Germany on good soil, but it turns out that miscanthus plantations are labour-intensive. Furthermore, the plant does not react well to bad weather. At present, the hype has largely died down around miscanthus. Recent research tends to focus more on grain, corn and fast-growing trees and shrubs.[19]

Low energy density places one other general limit on energy crops. For biomass to become competitive, transport costs have to be low. That generally means that plantations have to be relatively close to consumers. Long-range transport of biomass generally only makes sense if an energy-rich concentrate, such as oil or biofuels, is made out of the original biomass.

High oil prices in recent years have brought more attention to energy crops. The negative effects of such plantations has also come into the foreground, which is why energy crops are limited both in Germany and in the EU (see 5.10):

- The environmental impact from energy crops can be even greater than the impact from conventional agriculture. For instance, if corn is planted as an energy crop, it impacts the regional water balance and leads to greater soil erosion than plantations of green wood. And if large monocultures of energy plants are planted on unused farmland, biodiversity is reduced because there are fewer wild areas.
- In Germany and the EU, energy crops directly compete with crops for food production. Whatever brings in the most money is planted. As oil prices rise, the pressure to replace food crops with energy crops will only increase. As a result, more food will have to be imported. Our demand for energy will then conflict with demand for food among the poor in developing and industrializing countries. We hardly need to ask whether a well-to-do Mercedes driver or a Bolivian miner will win out.

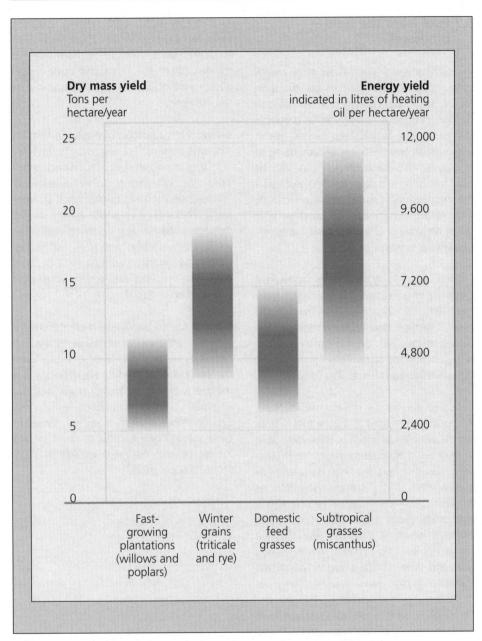

Figure 5.6 The potential energy yield of different energy crops in Germany

Source: Lewandowski and Kaltschmitt, 1998

5.7 Fuel from the field – biodiesel

In Germany, roughly a quarter of all energy consumption takes place in the transport sector. Our solar energy supply offers a number of options. Biogas can be produced and used as a fuel just as natural gas is. Entire urban bus fleets are already doing so in Sweden.[20] Solar hydrogen can also be used in mobile fuel cells (see 8.5). But by far the most common option today in Germany is biodiesel, which can be used without major revamping of vehicles and infrastructure (petrol stations, etc.).[21]

Biodiesel is generally made from rapeseed in Germany, which is currently planted on more than 700,000 hectares.[22] The fruit of the plant is first pressed to extract the oil. The 'rapeseed cake' left behind can be used to feed animals. The straw from the rest of the plant is returned to the soil as fertilizer.

Rapeseed oil (similar to canola in North America) can be used as a pure fuel in refitted diesel engines, which is commonly done in trucks and agricultural vehicles.[23] If rapeseed oil is esterified, the result is rapeseed-oil methyl ester (RME), commonly known as biodiesel. This renewable fuel can be used in almost any diesel engine, the most prominent of which is probably the German parliament building in Berlin, which receives heat and power from a cogeneration unit that runs on biodiesel.

Up to 2006, fuels made from biomass were exempt from the tax on mineral oil in Germany. Politicians did so to provide an incentive for renewable fuels.[24] As a result of this tax exemption and rising oil prices (themselves partly due to greater taxation of gasoline and diesel), biodiesel sales boomed in Germany. Plantations of rapeseed became a common sight, and refinery capacity exceeded 2 million tons in 2006. From 1998 to 2005, sales skyrocketed from 0.1–1.8 million tons.[25] But this trend came to an abrupt end in 2006, when biodiesel lost its tax exemption.

Biodiesel has a number of positive effects on the environment. For instance, finite fossil resources are offset, and sulfur dioxide emissions are also lower. Unlike mineral oil, biodiesel decomposes quickly in the environment without causing any environmental damage. Biodiesel also performs well when it comes to carbon emissions, which are roughly two thirds lower than those of diesel – a figure that rises even further when pure rapeseed oil is used.[26]

Roughly half of the rapeseed plantations for biodiesel are on land set aside by the EU. Farmers can thus decide whether to leave the land fallow or produce energy crops. But there is a ceiling on energy crops, and the allotment has almost been exhausted. Experts believe that biodiesel made in Germany will only be able to cover around 70 per cent of the diesel market by 2015 (2005: 5.5 per cent).[27]

90

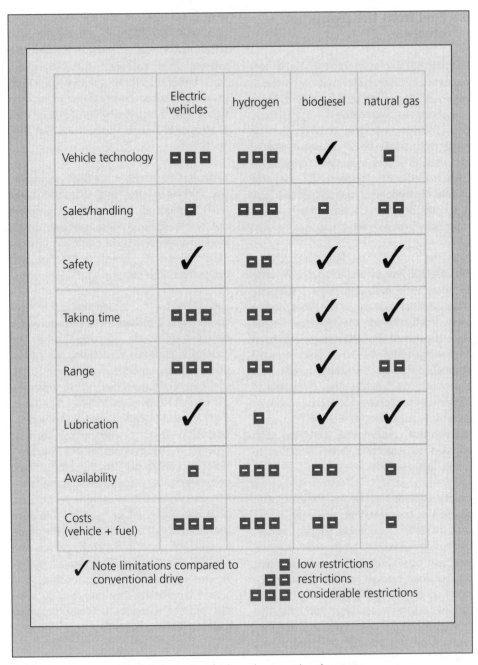

Figure 5.7 A comparison of alternative fuels and conventional systems

Source: Authors' depiction based on Shell PKW-Szenarien, 1998

5.8 Fuel from the plantation – ethanol

Roughly 51 per cent of the liquid fuel consumed in Germany for transport is diesel, with gasoline making up 45 per cent. There is a renewable alternative for gasoline as well. Ethanol (industrial alcohol) can be made from sugar beets or wheat (starch) and used as a fuel in gasoline engines. In Germany, ethanol is hardly used (3 per cent of fuel consumption, mainly blended in[28]), but Brazil and the US have much longer experience with alcohol as fuel.

After the first oil crisis, Brazil launched its PROÁLCOOL Programme in 1975.[29] Because oil prices were high and sugar prices low, large plantations of sugarcane were used to produce ethanol for the transport sector. Since the end of the 1980s, Brazil has been producing some 11–13 billion litres of ethanol per year. In 2005, production increased to 16.5 billion litres, equivalent to 45 per cent of global ethanol production. At present, Brazil covers 12–15 per cent of its national fuel consumption with ethanol in addition to exporting large amounts to China, Japan and the US.

Initially, ethanol was used in a mixture with gasoline at concentrations of up to 25 per cent in Brazil. At such concentrations, engines do not have to be changed. In 1979, the first cars running on 100 per cent ethanol went on sale, and they have since made up as much as 90 per cent of all new car sales in some years. Since 2003, flex-fuel vehicles (FSV) have been made in Brazil; they can run on any mixture of gasoline and ethanol.[30]

Brazilian ethanol from sugarcane has a clearly positive energy balance and impact on the climate. The fuel contains more than eight times more energy than was invested in it, and 150–220kg of CO_2-equivalent is offset per ton of sugarcane through the substitution of gasoline.[31] These values are much better than for the ethanol made from grain and sugar beets in Germany (see Figure 5.8), which is why Brazil exports a lot of ethanol to Europa – some 1.4 billion litres in 2008.[32]

To promote ethanol, Brazil does not charge any mineral oil tax, and value-added tax (VAT) is reduced. Middle class car drivers mainly benefit from these subsidies, which amount to around US$2 billion per year. At 2007 oil prices, Brazilian ethanol was competitive with petrol.

Sugarcane for the production of ethanol covers some 2 million hectares, roughly 3.5 per cent of Brazil's agricultural land. The country is currently considering an expansion of ethanol production to cover 10 per cent of all agricultural land. While sugarcane can easily be grown on underused pastures, there are also plans to have sugarcane plantations in sensitive ecosystems, such as swamps in flood zones in the Pantanal. Here, the general problem that biofuels face becomes clear: if a considerable share of the fuel we consume is to come from biomass, large plantations will be needed.[33] The extra land needed will either have to come at the expense of food production, which will detrimentally affect countries already suffering from malnutrition, or monocultures will reduce biodiversity. The only way to solve this conflict is to reduce fuel consumption by traveling less and becoming more efficient.

5

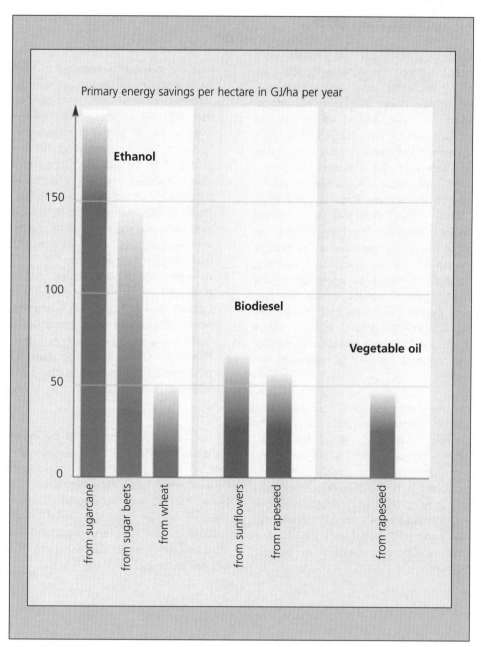

Figure 5.8 The energy payback of various fuels

Source: Authors' depiction based on Ifeu, 2005

5.9 Synthetic fuels (BTL)

Synthetic fuels made from biomass are a recent development. At present, there are only research and pilot systems. Nonetheless, hopes are high for biomass-to-liquid (BTL, also known as synfuel and sunfuel) fuels. A wide range of biomass – from wood to straw and other energy crops – is converted into a synthetic gas, from which methanol or a fuel similar to diesel is made.[34] The process is similar to Fischer-Tropsch synthesis. Such procedures are well known in the chemical industry. What is new about BTL production is the use of biomass as the precursor; a number of problems therefore still have to be solved. On the one hand, vegetable material is more heterogenous than oil and natural gas; on the other, the low energy density and distributed production of biomass mean that storage and transport have to be optimized (see 5.6).

The largest pilot unit for BTL fuel (Sundiesel) was inaugurated in 2008 in Freiberg, Germany. The plant, owned by Choren, which is partnered with car companies VW and Daimler and the Shell oil company, produces BTL fuel from any kind of biomass, such as wood chips, straw, weeds or leftover milk rejected by the agrofood industry. The plan is to have the plant produce 18 million litres of Choren's Carbo-V (a type of biodiesel).

Two specific benefits are expected from BTL fuel:[35]

1 Cars are increasingly expected to have lower emissions and better gas mileage. Such engines need fuels with properties that fall into an increasingly narrow window. The properties of rapeseed oil fluctuate slightly depending on the quality of the material used, so it seems clear that rapeseed oil will not be able to comply with the Euro 5 standard. But the hope is that synthetic fuels will be tailored to the requirements of specific engines in order to reduce emissions and make engines more efficient.

2 A wider range of plants can be used to make BTL fuels than to make other biofuels (see 5.7 and 5.8),[36] and the entire plant can be used, not only the fruit. The energy yield is thus higher. Nearly 4000 litres of fuel can be produced from a hectare, roughly twice as much as with biodiesel (see Figure 5.9).

On the other hand, the use of the entire plant also has a drawback; because no nutrients are returned to the soil, fertilizer has to be used, which raises the energy input. The humic content and overall productivity of the soil can be drastically altered. Here, sustainable agriculture is a must, and we must be careful not to quickly assume that BTL will be the biofuel of the future.

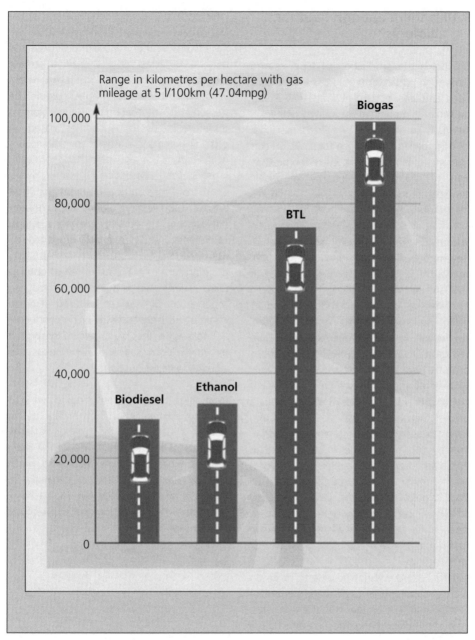

Range in kilometres per hectare with gas mileage at 5 l/100km (47.04mpg)

Figure 5.9 How much a hectare can produce

Source: Spiegel Special 5/2006, authors' depiction

5.10 Is there enough land for biofuels?

Expanding the production of biomass and biofuels in particular would require tremendous amounts of land – land that is not available in unlimited quantities either in Germany or elsewhere. At the same time, land is needed to produce food. As greater amounts of land are devoted to the production of biomass for energy purposes, conflicts are therefore inevitable. The figures below make that clear:

Worldwide, some 50 million square kilometres were available for agriculture in 2000, equivalent to 8200m² per person. Most of this land is used extensively, however, such as grasslands in Argentina. Only around 15 million square kilometres, roughly 2500m² per person, is used intensively. This figure will drop, however, as growing populations overtake production increases. By 2030, only 1900m² of farmland will be available per person.[37] Residents of EU-15 already make a greater than average use of their land and even import food and animal fodder, thereby taking up large tracts of land in Africa and Latin America.[38] Imports of biofuels to fulfill the EU's 20 per cent target by 2020 would increase the land needed by 17–38 per cent, roughly bringing us up to three times the area available per person (see Figure 5.10).[39]

Clearly, the EU will not be able to make do with its own land and will therefore require land abroad. The danger is that energy crop plantations will expand intensive agricultural land, leading to further clear-cutting of rainforest as can be seen today in Brazil and Malaysia, for example.[40] Furthermore, there is a danger that these countries will rededicate land previously used for domestic need in order to produce biomass for export, which would endanger the livelihood of the local rural population. At the same time, we have to ensure that the use of pesticides and mineral fertilizers is not used to create monocultures in order to increase biomass yield. The ecological impact would be tremendous. Internationally traded biomass therefore needs to be certified based on ecological and social aspects to take account of these risks.

Crop rotations and combinations that ensure biodiversity along with greater harvests provide opportunities for a supply of biomass,[41] and new energy crops that do not change ecosystems much can also play an important role. For example, jatropha, a plant usually used as a hedge in Africa, can produce an oil that can be used in diesel engines. Tests are currently being conducted in India with the plant, which grows on extremely poor soil and nonetheless has significant oil production.

In addition to our current mobility, we also have to take a closer look at our eating habits. Meat production requires large amounts of fodder and is one of the main reasons why EU citizens require so much land. By changing our diets, we can make do with less land. Otherwise, we face a bleak alternative – steak on your plate or biofuels in your tank.[42]

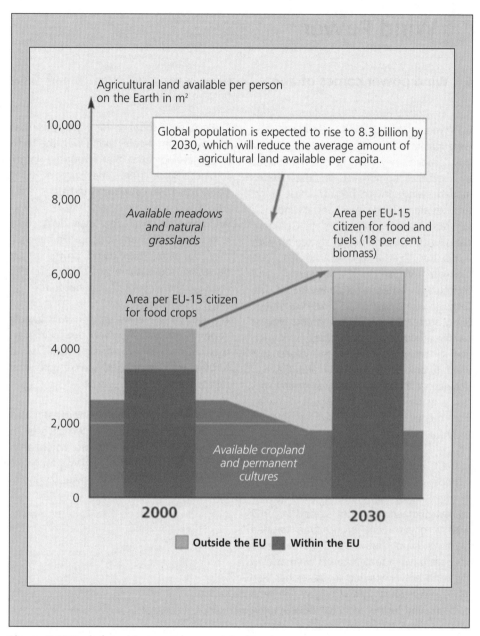

Agricultural land available per person
on the Earth in m²

10,000

Global population is expected to rise to 8.3 billion by
2030, which will reduce the average amount of
agricultural land available per capita.

8,000

*Available meadows
and natural
grasslands*

Area per EU-15
citizen for food and
fuels (18 per cent
biomass)

6,000

Area per EU-15 citizen
for food crops

4,000

2,000

*Available cropland
and permanent
cultures*

0

2000

2030

Outside the EU   Within the EU

Figure 5.10 Fuels from biomass take up a lot of land

Source: Authors' depiction based on Bringezu and Steyer, 2005

6 Wind Power

6.1 Wind power comes of age

Back in 1900, some 18,000 traditional wind-mills were still in operation in Germany. They were mainly used to produce flour. But over the decades, they were gradually replaced by electrical equipment. In the 1980s, modern wind power began. These wind turbines are no longer mechanical windmills, but rather power generators. In Germany, the new age of wind power started off with a disaster. In the Growian project, the German Ministry of Research and Technology wanted to set a new record. The Growian wind turbine had a rated output of 3MW, some 55 times larger than the biggest serially produced wind turbines back then and six times bigger than the largest test wind turbine in Denmark – and German engineers did not have nearly as much experience as their Danish colleagues. Growian was quickly decommissioned because of technical problems.

In the 1980s, Denmark retained its pioneering position in the field of wind power, but in the 1990s the German wind power market boomed. It all got started in 1989 with a government programme entitled '250MW Wind', but in 1990 the ball really got rolling when feed-in tariffs were offered for wind power starting in December (see 11.9). Two-figure growth rates became common and further technical development produced ever larger wind turbines. At the end of the 1980s, 50kW was a large machine, and 250–300kW was state-of-the-art at the beginning of the 1990s. Nowadays, large wind turbines generally have outputs exceeding 1MW, and 2MW is nothing unusual. The largest standard turbines on offer have 5MW, and tests are currently being conducted on even larger units.

The economic situation for wind power has improved considerably along with the technical developments that produced larger turbines. Since 1990, the price of wind power has been cut roughly in half.[1]

And although units larger than 2MW could not currently offer any specific cost benefits, they do allow more wind power to be harvested from a given plot of land, thereby reducing the amount of land needed.

The future of wind power will largely depend on two further developments: repowering i.e. replacing small wind turbines with larger ones; and offshore wind farms (see 6.7).

The potential for onshore wind is estimated at 45–65 terawatt-hours (TWh) per year in Germany alone. The offshore potential in Germany is estimated at 110TWh, roughly a quarter of annual German power consumption.[2]

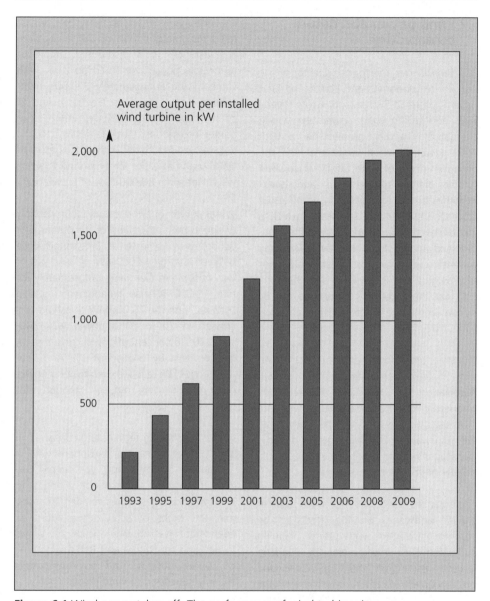

Figure 6.1 Wind power takes off: The performance of wind turbines increases

Source: The authors; DEWI

6.2 Wind power and nature conservation

H.C. Binswanger, the former director of the Institute of Economics and Ecology at Saint Gallen College in Switzerland, wrote several years ago calling wind power 'the wrong alternative'. The main problem he has with wind power is 'visual emissions' and the negative impact of wind farms 'when wind turbines are a blight on entire landscapes'. He writes that he is 'mainly concerned about protecting our cultural landscape, which is predominantly characterized by multifarious farmland and pastures that arrived at their current unmistakable form over long periods of slow growth'.[3] Binswanger says there is a risk that these cultural landscapes will be turned into large tracts of 'technological parks'.

The Swiss professor is not alone in opposing wind power.[4] For instance, the German Association for Landscape Protection is frequently on-site whenever there are protests against a wind turbine. They agree with Binswanger that wind power destroys landscapes without making any considerable difference in our energy supply.

There can be no doubt that wind turbines change landscapes. But the effects can be kept to a minimum with proper planning and citing. Furthermore, relatively little land is used.[5] Aside from the foundations (and possibly access roads and transformers), the soil remains unchanged and can still be used in agriculture.

And of course, taste is a matter of opinion. While wind turbines may be a thorn in the side of some people, others may see them as fascinating symbols of progress. More importantly, opponents of wind power should say why they do not oppose our conventional energy supply system, which has a far greater impact on our landscapes. For instance, in a distributed supply of renewable power, we could do away with some of the approximately 200,000 high-voltage power pylons in Germany. And of course, coal mining destroys entire landscapes into moonscapes and destroys towns. Yet, a wind turbine with an output of 1.5MW and a service life of around 20 years will offset some 80,000 tons of brown coal.[6]

Wind power opponents are demonstrably wrong in their estimation of the potential of wind power. For instance, Binswanger once believed that only 5000MW of wind power was possible in Germany; unfortunately for him, 10,000MW had already been installed by 2002, and the 25,000MW threshold was crossed in 2009. Wind power will cover some 20–30 per cent of power consumption in Germany by around 2025. And that target could be reached even faster if power consumption were reduced through efficiency and conservation.

If your goal is long-term nature conservation and protection of the biosphere with a sustainable energy supply, you simply may have to accept some minor changes to cultural landscapes. And if you can, then you join the ranks of such organizations as Friends of the Earth, Greenpeace, the World Wide Fund for Nature and Robin Wood, all of whom support the environmentally friendly expansion of wind power.

At a good location, a 1.5MW wind turbine with a hub height of 67m will produce some 76 million kilowatt-hours of electricity over 20 years of operation. To produce the same amount of power in a modern brown coal plant, some 84,000 tons of brown coal would have to be fired. If that amount of coal were piled up, the pile would be 50m tall and have a diameter at the base of 80m.

Figure 6.2 Wind power and landscape conservation

Source: The authors

6.3 Wind velocity is key

Wind turbines have basically four operational phases. When the wind is not blowing enough to overcome the turbine's frictional resistance, the turbine does not move. When wind velocities reach around 3m per second (around 10km per hour), the turbine begins to turn and generate electricity. As wind velocities increase, the turbine's output increases until it reaches the generator's nominal output – the maximum amount of power it can produce. As wind velocities increase even further, the excess power has to be done away with. Generally, the rotor wings are either specially designed to do so, or they can be pitched out of the wind mechanically. Finally, in storms (wind velocities of 20–25m per second upwards), the turbine switches off automatically to prevent damage.

The average amount of power produced by wind turbine installed in Germany is around 1800kWh per kilowatt of installed capacity.[7] Regionally, though, values differ. On the coast of the North Sea, power output is roughly 13 per cent greater than the German average, while it is 10 per cent below average in Rhineland-Palatinate. Wind turbines at high altitudes can, however, perform as well as turbines on the coast.

The main reason for different annual performances is different wind velocities and frequencies from one region to another. Average wind velocity is the main factor in a wind turbine's output. Generally, greater wind velocity increases the power output by a power of three.

In other words, if the output of the wind turbine is 100 per cent at 10m per second, that turbine does not produce 10 per cent more power under a wind velocity of 11m per second, but instead roughly 33 per cent more – $(1.1)^3 = 1331$. Put differently, 10 per cent more wind velocity means a third more wind power.

When siting wind turbines in hilly regions, this fact is crucial. Often, nature conservation proponents claim that a wind turbine on top of a hill would be a blight on the landscape; they therefore require the turbine to be located on a slope. Yet, wind velocities are generally higher at peaks. The difference for wind power can be severe. For instance, if an average wind velocity at the peak is 6.5m per second over the year compared to 4.5m per second on a slope lower down, the difference is around 30 per cent, which may not seem crucial to a layperson. But, in fact, the wind turbine will produce more than 65 per cent less power, making the turbine lower down the hill unprofitable. And, if you want to build it anyway, you would need to put up three turbines on the slope to get the same power that a single turbine on top of the hill would produce. On the one hand, that option is an inefficient use of resources; on the other, defenders of landscapes should decide whether they want to have a single turbine in a good location or three turbines on suboptimal locations.

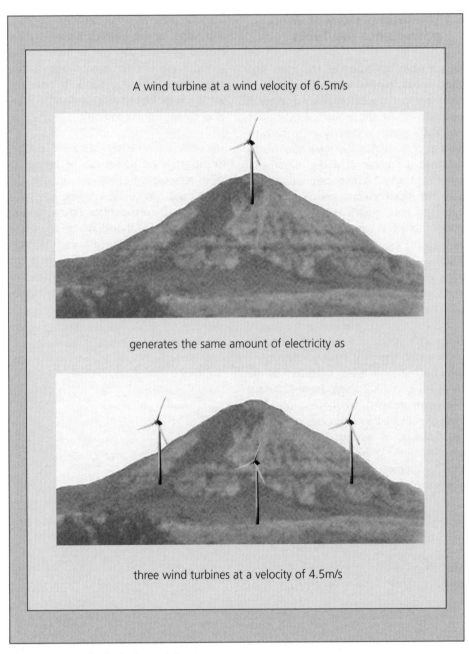

Figure 6.3 Wind velocity is crucial

Source: The authors

6.4 The success story of wind power since the 1990s

To see what opportunities the Solar Age offers to industry and trades, we need look no further than the history of wind power in Denmark. While the wind energy market had hardly gotten underway in Germany by the end of the 1980s, Denmark had already instituted a number of policy instruments with clear targets.[8] A dedicated research and development programme along with feed-in rates and clear guidelines to export wind power to the grid allowed a new market to be created. As a result, a booming wind industry was set up in Denmark within only a few years. The Danish industry became the world's main manufacturer and exporter of wind turbines.

With some delay, Germany became the country with the most installed capacity of wind turbines until it was overtaken by the US in 2008. In 2002 alone, some 3200MW was newly installed. By the end of 2009, more than 21,000 wind turbines with a total rated output of around 25,700MW had been installed. In 2004, wind power overtook hydropower as the largest source of renewable energy in Germany. [9]

Manufacturers of wind turbines, suppliers, and wind power project planners have created some 85,000 jobs in the past few years,[10] many of them in small and midsize enterprises. The wind sector is now the second largest purchaser of German steel after the automotive industry.[11]

The positive experience gained with policies to support wind power can be applied to other renewable technologies, though the focus need not be on applying the same rates with the exact same policy designs. Rather, the goal should be to develop a mixture of policies tailored to each technology and field of application so that the current roadblocks in the implementation of efficient, environmentally friendly technologies can be overcome (see 11.18).

6

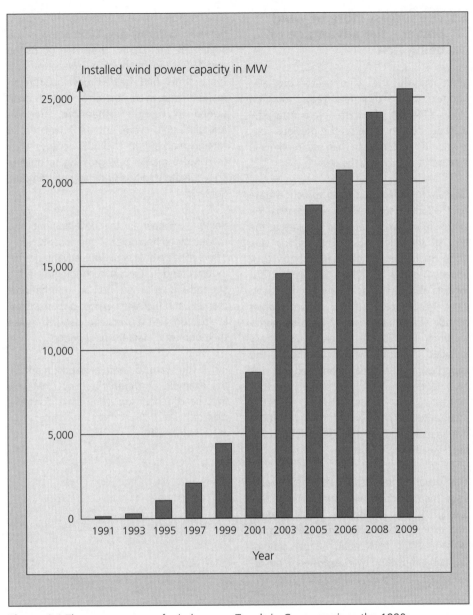

Figure 6.4 The success story of wind power: Trends in Germany since the 1990s

Source: The authors; DEWI

6.5 The success story of wind power – the advantage of being first

Internationally, the wind sector has also boomed in the past few years. In 2004 alone, 8154MW of turbines were installed,[12] a 20 per cent increase on the previous year. In addition to European markets, markets in Asia and the US grew the fastest.

German manufacturers used to concentrate on their domestic market, which was the largest in the world for wind turbines at the turn of the millennium. They had a very good market position back then; they installed some 2100MW in 2001. But exports were few and far between for them. Only 14 per cent of the global market outside of Germany went to German manufacturers (3805MW). In terms of newly installed capacity within Germany, the export quota of German manufacturers was 16.4 per cent in 2001.[13] In contrast, the four biggest Danish manufacturers exported around 3000MW of wind turbines in 2001, five times as much as their German competitors[14] (see Figure 6.5).

The situation has changed considerably since then. German wind turbine manufacturers are now among the global leaders. And exports make up a large part of their sales. In 2005, more than 60 per cent of German production was exported,[15] ensuring approximately 30,000–40,000 jobs.

One reason for the disparity in exports between Germany and Denmark was the advantage that Danish firms had up to the mid-1990s in technology development; these firms had helped develop foreign markets and were therefore better positioned in them. Furthermore, German manufacturers were not as interested in foreign markets up to 2003 because their own home market was growing so quickly. They therefore focused on meeting domestic demand.

Being a pioneer in the development and rollout of technologies – the first mover – offers considerable economic advantages for industry and trades. A technological edge strengthens your position in international markets and increases export opportunities. In addition, first movers can also help make their domestic markets independent.

But if you want to conquer export markets, the example of Denmark shows that you first need to set up your own domestic market.

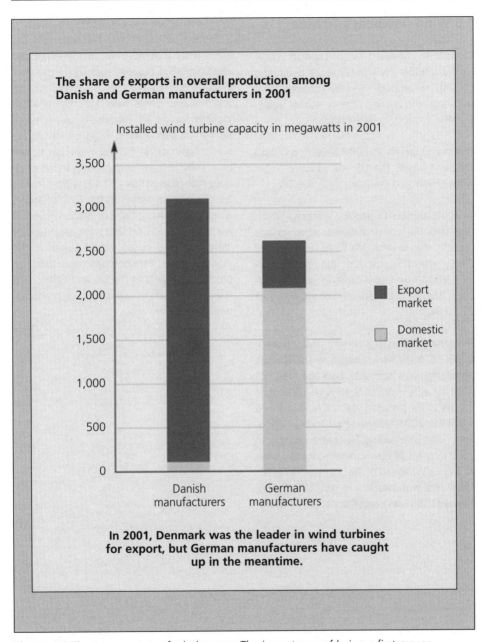

Figure 6.5 The success story of wind power: The importance of being a first mover

Source: Authors' depiction based on DEWI

6.6 Wind power worldwide

Worldwide, installed wind capacity grew from just under 20,000MW at the beginning of 2001 to around 158,000 megawatts at the end of 2009. Within eight years, installed capacity grew eightfold.[16]

The main markets in 2009 were China with around 13 GW, the US with 10GW, Spain with 2,5GW and Germany with 1,9GW.

In 2008, the US overtook Germany, which had had the most installed wind power capacity for several years. At the end of 2009, the US still led the pack with 35,170MW. Germany came in second with 25,770MW just ahead of China at 25,104MW.[17]

The greatest growth is currently taking place in the US and China. Thanks to the Obama administration's stimulus package, the US wind industry installed 9.9GW in 2009 after 8.4GW in the previous year.[18] China installed 6.3GW in 2008 followed by 13GW in 2009. Spain was the leading European country in 2009 in terms of newly installed wind power at 2459MW, ahead of Germany (1917MW). Much less was invested in Italy (1114MW), France (1088 MW) and the UK (1077 MW).

Wind power has also become big business in a number of emerging nations, such as India. After a number of already impressive years, installed capacity rose by 23 per cent in 2008 and 13 per cent in 2009. Most of the demand comes from industry, which is looking for an alternative to conventional power suppliers. Here, high oil prices are also a main driver behind the wind power boom. As long as the price of a barrel of oil does not drop below US$40, wind turbines are cheaper according to the CEO of Indian wind turbine manufacturer Suzlon.[19] Suzlon is a solely Indian firm that has grown quickly along with the Indian wind market. In the past few years, the company has also set up production plants in the US and China, and overtaken German wind turbine manufacturer Repower.

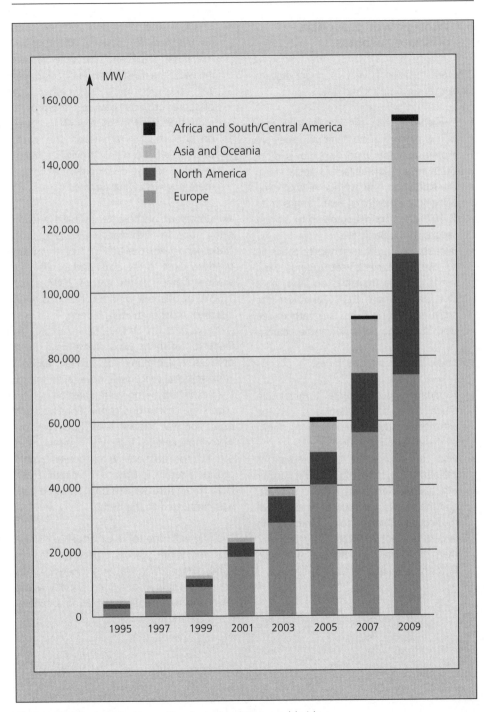

Figure 6.6 The installed capacity of wind turbines worldwide

Source: World Wind Energy Association

6.7 Wind power prospects – offshore turbines

Offshore turbines have two major advantages over onshore technology:

- The energy yield per installed capacity unit is much higher. Wind velocities are higher, and the wind also blows more often because of the lack of obstacles. In the North Sea, the average annual wind velocity at a height of 60m ranges from 7–10m per second, compared to 8m per second in the Baltic Sea.
- Because the area is generally available for use, it is easier to set up large wind farms offshore than it is on land. At a distance of more than 10km off the coast, such wind farms are hard to see from shore and no noise can be heard.

But there are also disadvantages:

- Foundations for offshore turbines are much more complicated and can be extremely costly depending on the depth of the water (up to 40m).
- Grid connections for such wind farms are also a major cost factor. The further the turbines are away from the coast, and the smaller the wind farm, the more these costs will play a role. Generally, the largest turbines available (usually larger than 2MW) are used offshore, and such farms have a large number of turbines.

- Finally, maintenance and servicing costs are still hard to calculate. The turbines run more often and the salty sea air increases the need for anti-corrosion. We will, however, soon know whether such offshore projects pay for themselves more than wind farms in good locations on the coast.[20] Whatever the case, offshore wind certainly provides a major incentive for the development of larger, more powerful wind turbines.

A number of offshore farms have already gone up in Denmark, Sweden, The Netherlands and the UK.[21] All of them are relatively close to the coast and in relatively shallow water.[22] By the end of 2008, some 1.5GW of offshore wind turbines had been installed, most of them in Europe.

Plans for offshore wind farms have been delayed in Germany for several reasons. First, approval procedures have taken longer than expected. Second, the question of who covers the cost of grid connections and grid expansion was not yet clear. Third, investors were not certain that the feed-in rates offered for offshore wind power would provide a return in light of the higher costs. In 2010 the first German offshore wind park was connected to the grid.

The grid will have to be expanded considerably for wind power, especially offshore wind. Germany's Network Agency has produced a study[23] on what Germany would need and is currently completing a follow-up.[24]

Wind power in shallow seawater:

great expenses, but also great returns

Figure 6.7 Tremendous potential for offshore wind

Source: VESTAS

6.8 Wind power prospects – less is more through repowering

Repowering means replacing small, old turbines with new, more powerful and more efficient ones.

Wind turbines have an expected service life of 20 years, but the technology has developed so much over the past two decades that wind turbines can seem outdated even though they may still be running. For instance, in 1995 the average turbine newly installed in Germany had an output of less than 500kW; by 2008, it was around 2MW. Larger turbines with taller towers could make better use of wind and run at higher capacity. The energy yield of modern wind turbines has thus been growing faster than their rated capacity. As a result, it would be possible to produce even more power within a given plot of land if old turbines are replaced by larger – and more efficient – new ones. The rule of thumb is 'twice the output from half as many turbines'.

The greater power yield comes from the turbine's greater output (longer, more efficient wings, higher nacelles) and from greater availability.

Repowering is a political goal. Since the German Renewable Energy Act was revised in 2004, special financial incentives have been provided to replace inefficient old turbines that are nonetheless still running properly.

However, a number of obstacles prevent or slow down repowering. For instance, the State of Lower Saxony recommends that its communities only install new turbines at least 1000m from the nearest home.[25] Other German states have similar stipulations. Furthermore, some communities do not allow turbines taller than 100m to be set up, which practically rules out turbines with 2MW or more.

Such height and distance limits not only severely restrict the expansion of wind energy, but also ensure that the impact on the landscape is greater than necessary. After all, repowering allows the number of turbines installed to be reduced. For instance, in a project in northern Germany (see Figure 6.8) three five-megawatt turbines could replace eleven 500kW machines[26] even as the overall annual output is tripled.

Repowering has been going slowly in Germany. In 2005, only six wind turbines were renewed. The revised Renewable Energy Act of October 2008 provided better financial terms for repowering. The success of this new law is already clear. While some 24MW was repowered in 2008, the figure had risen to 136MW in 2009. As the President of the German Wind Energy Association (BWE) put it: 'Germany has a lot of wind energy potential in repowering over the next few years.'[27]

Simonsberg wind farm	Before	After	Multiplier
Number of turbines	11	3	0.27
Hub height (x metres)	42m	120m	2.86
Rated output of individual turbines	500kW	5,000kW	10.0
Total installed capacity	5.5MW	15MW	2.72
Full capacity hours	2,545	3,200h/a	1.26
Annual power production	14 Million kWh	48 Million kWh	3.43

Figure 6.8 Repowering: Less is more

Source: Bundersverband Windenergie, 2005

7 Water Power, Geothermal and Other Perspectives

7.1 Water power – the largest source of renewable energy

Solar energy keeps water circulating on the Earth. Worldwide, more than 50 billion cubic metres of water evaporates each hour, most of it over the oceans. When this water falls onto land as precipitation, the difference in altitude between the land and sea level provides useful energy potential.

Water power is one of the oldest sources of energy used by mankind. In around 3500 BC, water mills were used in Mesopotamia to irrigate the land. And around 100 BC, water mills were used to grind grain and drive sawmills in the Roman Empire. Today, the kinetic energy in water is mainly used to generate electricity. The process is as easy as it is efficient; water flowing downhill drives a waterwheel or a turbine, which generates electricity. Depending on the amount of water and the difference in altitude, different types of turbines are used,[1] with the average efficiency being quite high at 90 per cent. Hydropower has one advantage over wind power and photovoltaics: it is relatively constant and can be stored easily. As a result, a number of hydropower plants can be used to provide baseload power. And water stored in lakes can also be sent through turbines in seconds to provide peak power. Water power is therefore a very flexible source of renewable energy.

Worldwide, hydropower is by far the largest source of renewable electricity. In 2005, some 2950TWh of hydropower was generated, equivalent to 90 per cent of the electricity from renewable energy and some 16 per cent of all of the electricity generated worldwide. Hydropower comes in just ahead of nuclear power, which provided 2771TWh in 2005. In some countries, hydropower is the largest source of domestic electricity, for example, in Canada (60 per cent), Brazil (84 per cent), Switzerland (around 55 per cent), Iceland (around 80 per cent) and Norway (around 98 per cent).[2] Iceland and Norway are also looking into ways to expand hydropower even further in order to provide renewable hydrogen from excess electricity; the hydrogen could then be used as a fuel in vehicles, for example (see 8.5).[3]

Yet, the potential of hydropower has hardly been exhausted. Worldwide, it is estimated that some 15,000TWh of electricity could come from hydropower each year, roughly equivalent to our current global power consumption, though only half of that is considered economically feasible. Unfortunately, the potential that hydropower has is not equally distributed across the globe (see Figure 7.1). Europe and the US already use a large part of their potential, whereas Africa, Asia and South America could still expand greatly.[4]

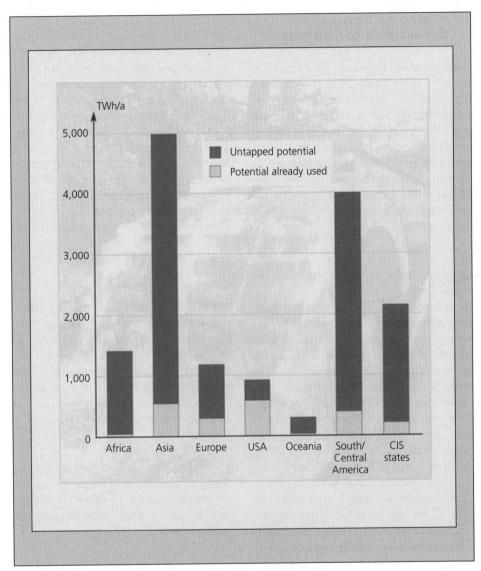

Figure 7.1 Water power: Global potential, used and untapped

Source: Landesinitiviative Zukunftenergien NRW: Wasserkraftnutzung

7.2 Expanding hydropower – the example of Germany

Hydropower has been used for centuries in Germany. Often, hydropower was the starting point for merchants and cottage industries (mills, pumps, etc.). Starting in the mid-19th century, hydropower was used to generate electricity, and up to 2003 it was the largest source of renewable electricity in Germany. Today, hydropower provides some 20 billion kilowatt-hours of electricity per year, roughly 4 per cent of the country's electricity consumption.

At the same time, hydropower is the only renewable source of energy that has been largely exhausted in Germany; roughly 75 per cent of its potential is already exploited. The potential for new large hydro dams has already been completely used up, though current systems can generally be revamped to increase power output considerably.[5] For instance, the Rheinfelden plant was modernized to increase power production more than threefold. The power increased from 26MW to 100MW, and the annual power production, which began in 2010, will grow from 185GWh to 600GWh.[6]

Micro hydropower units, in contrast, still have a lot of potential. The German Hydropower Association (BDW) of Munich estimates that units with an output of up to 5MW could increase the amount of electricity from hydropower by around 50 per cent.

Hydropower potential is unevenly distributed across Germany because of the country's typology. The two southern German states of Bavaria and Baden-Württemberg have some 75 per cent of German hydropower potential (see Figure 7.2), while there is very little potential in the north.

Over the past 100 years, some 50,000 microhydro units have been decommissioned in Germany; in 1850, roughly 70,000 such units were still in operation. But often, they did not pay for themselves because the rates they received were too low, water rights were disputed or financing was hard to get for urgently needed repairs. But since the Feed-in Act of 1991, the situation is changing. From 1990 to 1999, the number of microhydro units that generate electricity increased from around 4400 to 5600.[7] These new systems alone generate some 80 million kilowatt-hours of environmentally friendly electricity, roughly enough to cover the power consumed by more than 200,000 German households.

Reactivating old units is not the only way to increase the amount of energy from hydropower. Technical improvements can also be made to old water wheels and turbines.[8] But generally, expanding and increasing the efficiency of hydropower units requires large upfront investments that only pay for themselves over long periods of time.

To expand hydropower further, current approval procedures, which are quite complicated, need to be simplified, and the right of use for water must be given for generous periods of time.

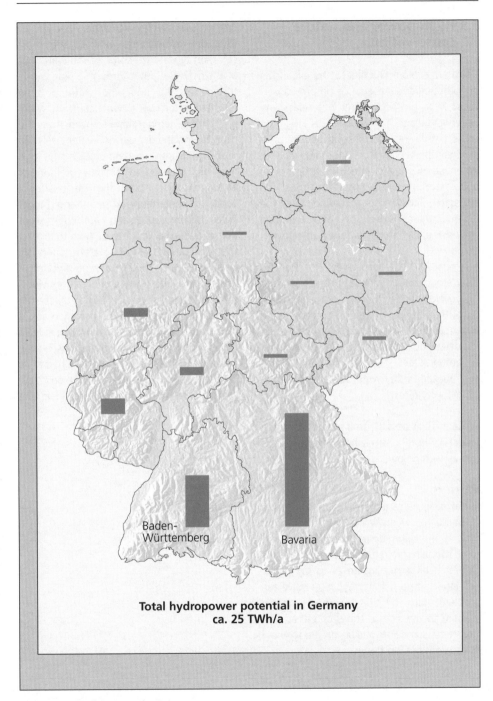

**Total hydropower potential in Germany
ca. 25 TWh/a**

Figure 7.2 Hydropower in Germany

Source: Kaltschmitt, Wiese. Chart: triolog

7.3 Hydropower and nature conservation

There are ecological limits to the expansion of hydropower. One goal of nature conservation is to protect natural waterways and waterways close to nature. This goal can conflict with the use of hydropower; after all, hydropower units change ecosystems. When permits are granted for hydropower units, the individual situation must be assessed exactly. The following must be taken into consideration and weighed against the advantages of renewable power:

- Impact on the ecosystems of flowing water, especially the protection and development of local flora and fauna both in the water and on banks.
- Impact on water management, especially flood protection, flow rates and groundwater.
- Impact on other water functions, such as recovery rates.

While it is not possible to discuss all of these aspects comprehensively here, we do discuss two aspects below:

Barriers

Run-of-river dams generally have some kind of barrier or weir that prevents fish from moving upstream easily, thereby stopping some species from returning to parts of their natural habitat (for instance, to lay eggs). Fish ladders and fishway bypasses solve the problem; part of the flowing water is directed to the side of the dam either over steps or in some sort of side stream to allow fish to continue up the river.

Minimum water volume

Diversion hydroplants, one of the most common types, take some of the water out of the river and put it through a pipe with a turbine at the end; the water is then returned to the river downstream. The water is therefore temporarily removed from the river. During the dry season, water levels can drop significantly, creating problems for local flora and fauna. The minimum amount of water that has to be in the river is often a bone of contention in permitting procedures. Nature conservation authorities want that minimum amount to be high to protect the local ecosystem, but plant operators want to be able to take a large amount of water out of the river to generate (environmentally friendly) power. For instance, the following compromise was recently reached for a project in Baden-Württemberg: at least a third of average minimum flow – the amount of water that the river still had at its lowest level on average over the past few years – over the year must remain in the river.

7

**Fish ladder
and
pool-and-weir stream**

Figure 7.3 Hydropower and nature conservation: Fish bypasses

Source: Das Wassertriebwerk, 1/2000

7.4 The world's largest hydropower plants

While Rheinfelden is considered a large hydropower plant by German standards, at only 100MW megawatts, it is quite small on a global scale. Worldwide, some hydropower plants are true giants.[9]

The Itaipú dam on the border between Brazil and Paraguay was constructed on the Iguaçu and Paraná Rivers. A storage lake some 170km long – roughly twice as large as Europe's Lake Constance – was created behind a dam 196m tall and 7.8km long. The power plant has a capacity of 12.6GW and generates around 95,000GWh of electricity each year, roughly enough to cover a quarter of Brazil's power demand – or theoretically a sixth of Germany's.

For a long time, this hydropower plant, which went into operation in 1983, was the largest in the world, but in May 2006, the Three Gorges Dam on the Yangtze River took that title. At 185m tall and 2.3km long, this dam is justifiably nicknamed the New Great Wall of China. Now that all of the turbines have been put into operation, the power plant has a capacity of 18.2GW, roughly as much as 16 nuclear power plants.

Despite the positive aspects of renewable electricity and flood prevention, these and other giant hydropower plants have faced severe criticism.[10] For example, the construction of such dams forces millions of people to leave their homes and resettle. At the Three Gorges Dam alone, an estimated 1.2 to 2 million people had to be relocated.[11] Generally, these people do not receive proper compensation for their losses. A number of them, such as fishermen, lose their means of livelihood altogether. The ecological impact of large storage dams is also criticized. The flooding of entire valleys completely destroys ecosystems, and the flooding of dense forests releases toxic gases.[12] Furthermore, the resulting lower levels of oxygen in the water detrimentally affect fisheries, and the new body of water affects the climate. For instance, the sediments carried by the Yangtze River – some 680 million tons of sand and mud each year – is gradually filling up the storage lake (which would fill up much quicker if it were not so deep), and the water below the dam lacks sufficient nutrients. In the tropics, storage dams are also a health risk because they provide optimal habitats for mosquitoes that transmit malaria.[13]

There were reports of widespread corruption during the construction of the Three Gorges Dam related to apparent fractures and holes in the dam itself – which, it should be noted, was constructed in a region prone to earthquakes that could destroy the dam altogether. If that happens, millions of people could be flooded.[14]

In conclusion, most large hydropower plants not only produce environmentally friendly renewable power, but also have a considerable social and ecological impact. These drawbacks must be taken into consideration when we work to expand hydropower. It may often be better to have several small projects done instead of one large one.

Figure 7.4 Giant hydropower: The Three Gorges Dam in China

Source: dpa

7.5 Geothermal worldwide

Unimaginable amounts of energy are stored within the Earth. More than 95 per cent of our planet is hotter than 1000°C. Heat is constantly flowing to the relatively thin crust of the Earth from its core. The amount of heat that reaches the surface worldwide is roughly equivalent to 2.5 times our global energy consumption. Up to now, most of it has escaped into space unused. But this geothermal heat can be an important part of our sustainable energy supply. After all, it is available around the clock in all seasons, unlike intermittent solar and wind energy. It can therefore be used to cover baseload demand, making it an important option to replace nuclear power.

Today, geothermal energy is used in 76 countries. Most of these systems are in regions with active volcanoes, where high temperatures are found not far below the surface. Geothermal energy can be used directly to heat buildings, greenhouses, swimming pools, etc. or as process heat to desiccate produce and fish, produce salt and for other purposes. Roughly half of the geothermal energy currently used comes from heat pumps, which generally run on electricity (see 7.6); some 2 million such heat pumps are currently used worldwide to heat and cool buildings.[15]

Geothermal heat can also be used to generate electricity. The first geothermal plant, which had an output of 250kW, went into operation in 1913 in Lardarello, Italy.[16] Today, geothermal power plants worldwide collectively have an output of around 9000MW, equivalent to 20 large coal plants. Most of them are in the US, especially in California (some 2540MW). In the world's largest field, the Geyser (1421MW), several power plants collectively produce as much electricity as a large nuclear plant.[17] Over the past few years, the Philippines, Indonesia and Japan have added a lot of geothermal electricity generating capacity, more than doubling their overall capacity from 1990 to 2000. In Java, Indonesia, Gunung Salak is installing one of the world's largest geothermal power plants (330MW).[18] Worldwide, geothermal power made up some 1.8 per cent of renewable electricity in 2003.

In Europe, Italy and Iceland are geothermal leaders. Iceland already gets 55 per cent of its primary energy consumption from geothermal energy; 87 per cent of homes there are supplied with geothermal heat, and greenhouses and fish farms heated with geothermal heat helped the country increase its food production.[19] Iceland also gets most of its electricity from hydro and geothermal plants. In 2004, some 200MW of geothermal power generation was installed, enough to cover around 15 per cent of the country's power supply. In 2008 the installed generation capacity of geothermal power plants totalled 575MW and the production was 4038GWh, or 24.5 per cent of the country's total electricity production.[20]

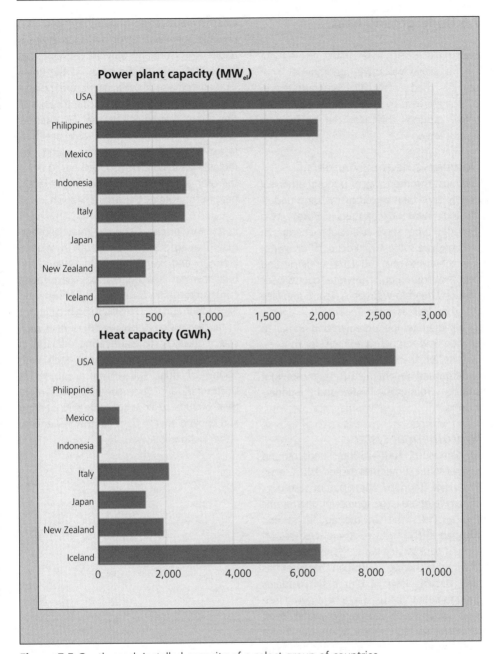

Figure 7.5 Geothermal: Installed capacity of a select group of countries

Source: Authors' depiction based on International Geothermal Association (IGA), 12/2004

7.6 Underground heat

Even in places such as Germany, where there are no active volcanoes, geothermal heat can be used. After all, temperatures in Europe increase by around 3°C every 100m below ground. This heat can be used in various ways:

Downhole heat exchangers

The most common way of using geothermal energy is to heat buildings by using underground heat exchangers. Here, two u-shaped tubes serve as a heat exchanger in the borehole some 100m deep. When water is injected into this tube, the underground heat is absorbed and the water comes back up a few degrees warmer. A heat pump (see 8.1) then increases the underground energy to a higher temperature (around 35°C) to provide low-temperature heat for homes. Outside of the heating season, the heat underground regenerates through sunlight and heat from ground water and the underground.

Hydrothermal systems

At depths of 1000–2500m, underground temperatures sometimes exceed 100°C even in parts of Germany. These natural reservoirs of hot water are large sources of geothermal energy. The water in the aquifer is first pumped to the surface, where it passes on its heat to a service water system by means of a heat exchanger. Although the water has now cooled down a bit, it still contains enough heat to be used in a heat pump (see 8.1), which cools down the water even further by raising the heat taken from it to a higher level. The water is then injected back into the aquifer. The heat collected can be used in a district heating network.

Geothermal heating stations are generally possible wherever there are such thermal aquifers. In Germany, they are mainly found in the flat lands of the north, in the Rhine Graben in the southwest, and in the molasse basin south of the Danube at the foothills of the Alps. According to the Geothermal Association of Germany (www.geothermie.de), the thermal aquifers along the molasse basin contain enough heat to cover the demand for environmentally friendly heat in three cities the size of Munich.

In the former East Germany, three geothermal heating stations went up in Waren, Prenzlau and Brandenburg. In Bavaria, the first thermal heating stations have been constructed in Straubing, Erding and Simbach-Braunau. The Straubing project, for instance, provides 21,600MWh of heat each year, equivalent to the heating demand in some 4000 low-energy homes each with 100m^2 of floor space.[21] In a project in Dortmund, an office complex with more than 6000m^2 of floor space gets its heating and cooling energy from geothermal energy – the Technorama.[22]

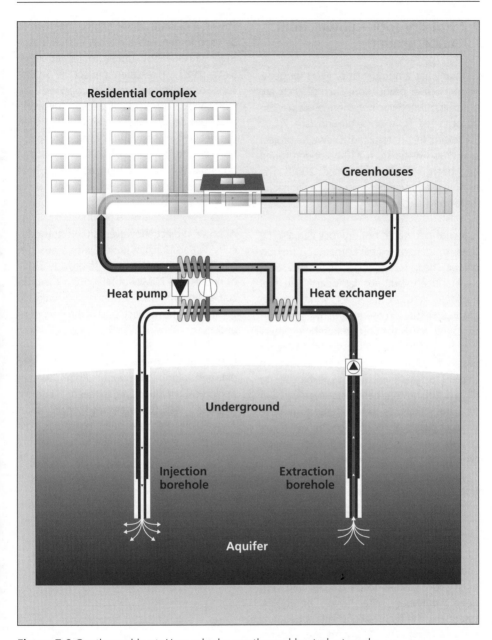

Figure 7.6 Geothermal heat: How a hydro geothermal heat plant works

Source: triolog

7.7 Hot dry rock – power from underground

Germany has a number of projects for geothermal power production. Such projects are promising wherever hot thermal water can be obtained from underground (see 7.6).[23] For example, in Neustadt-Glewe a project that draws water from 2200m underground has been in operation since 2003. The turbine used there generates 1.2 million kilowatt-hours each year.[24]

In lieu of hot water, hot dry rock can also be used to generate environmentally friendly power. The process involves boreholes some 5–7km deep, but not far apart. The first borehole is used to inject water underground. Fissures in the rock allow the water to flow towards the second borehole, where it is drawn back up to the surface. When it passes through the rock, it heats up so much that it creates steam, which can be used to drive a turbine that generates electricity. But first, the water has to be filtered to remove residual pieces of rock. Once the water has cooled down, it is pumped back underground into the injection borehole.

Hot dry rock (HDR) is possible if two things hold true. First, sufficiently hot layers of rock should not be too deep if the project is to be economically feasible. In Germany, this is the case especially in the upper Rhine Graben and in Upper Swabia. Second, the rocks must allow fissures to be created and kept open between the two boreholes. Essentially, the fissures constitute a geothermal heat exchanger. To create one, water is injected into the boreholes at high pressure. If the geological conditions are right, naturally present fissures are expanded to form a sort of connecting network. This network is crucial if the fissures are to stay open.

The HDR method allows geothermal heat to be used to generate electricity even in countries without hot underground aquifers. Since 1987, a research project in Soultz-sous-Forêts, France has been conducted on HDR. After many years of preparatory work, a 5000m deep, 1.5MW HDR system – then the deepest in the world – went into operation in the summer of 2008. A second borehole will expand the geothermal heat exchanger, bringing total output up to 3MW.[25]

Another project, however, shows the difficulty. In Bad Urach, a project with a hot rock of 170°C, 4.5km deep was to provide 3MW of power and 20MW of heat.[26] But the difficulties encountered during drilling were so great that the project has been discontinued.

Other geothermal power projects are planned along the upper Rhine, where geological conditions are excellent.[27] In most of these cases, however, hot thermal water is available. Because deep boreholes cost millions of euros, and there is no certainty that sufficient hot water will be found, the upfront risk is great in such projects.

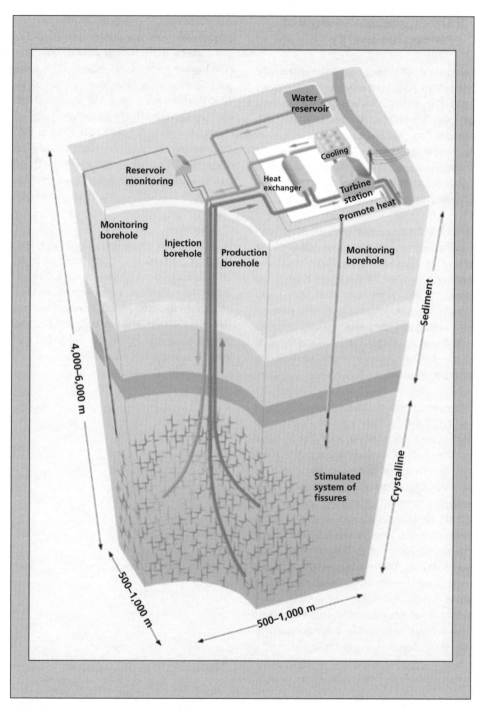

Figure 7.7 Power from hard rock: How the hot-dry-rock method works

Source: Markus O. Häring, Geothermal Explorer Ltd

7.8 Other possible sources of renewable energy

In previous chapters, we have presented types of renewable energy that have already proven useful in practice or are at least generally considered to be feasible going forward. Now, we will discuss a few types of renewable energy that are still in the test phase. This discussion cannot be exhaustive; rather, it is intended to show that new developments are possible with renewables.

For some time, people have been looking into ways to use the enormous power of the sea. From 1960 to 1967, a tidal power plant with a capacity of 240MW was built at the mouth of the Rance River near Saint-Malo, France. It takes advantage of the extreme tides using a dam and classic turbines. Unfortunately, there are only a few such locations worldwide that are suitable for such power plants.

British engineers in Devon, England, have come up with a new kind of tidal power plant. Essentially, they put a kind of windmill on the seafloor. Called Seaflow, the plant has already been tested at an output of 300kW.[28] Ocean waves also contain a lot of energy. The most promising approach to exporting this source of energy at the moment is the Pelamis project. Here, three interlinking steel tubes float on the water like a snake. The waves make the tubes bend against each other, and the motion drives hydraulics within the tubes, where power is generated. Some 750kW systems have already been sold and some new projects developed.[29]

Ocean energy has two clear advantages over wind power. First, the tides can be predicted and thereby make a more reliable contribution to our power supply. Second, flowing water contains far more energy than wind, which means that relatively small systems can produce relatively large amounts of energy. But there is a drawback to all of this: the great forces to be harnessed can also be destructive. If we are to use ocean energy on a large scale, we will therefore have to come up with a satisfactory return on our investment.

Solar chimneys are a very different project. Under a giant pane of glass, the sun heats up air, which rises very quickly up a tower in the middle. Turbines inside the tower then generate electricity. A pilot project with a 194m tall tower and an output of 50kW proved successful in Spain in the 1980s – at least technically, not economically.

Figure 7.8 (1) New power plant technologies: Solar chimneys

Source: Schlaich, Bergermann and Partner

Figure 7.8 (2) New power plant technologies: Seaflow

Source: ISET Kassel

8 New Energy Technologies

8.1 Heat pumps

Heat pumps take heat out of the earth, ground water or air and use them for heating purposes. They are therefore sold as environmentally friendly heating systems, especially by power providers. But is this true?

How they work

Heat pumps basically work like reverse refrigerators. While refrigerators take heat out of the inside and emit it on the outside (at the back), the pumps used in heating systems send the heat-carrying medium through a heat exchanger underground, for instance, to collect heat at a low temperature (around 10°C). The heat pump then brings that temperature up to around 35°C and sends the heat into a heating system. So while refrigerators create waste heat while cooling things down, heat pumps use the heat itself.

The heat can be used in various ways. If a geothermal collector is used, long tubes are installed some 2m deep in the earth; another possibility is a downhole heat exchanger (see Figure 8.1) extending down some 100m. But groundwater and ambient air can also be used. Because cold winter air does not contain much heat and groundwater is not available everywhere, most heat pumps in private homes use underground heat exchangers.[1]

The energy that a heat pump itself consumes increases the greater the temperature difference between the input (the temperature at the heat source, such as underground) and the output (the temperature that the heat pump provides to the heating system or hot water tank). Heat pump systems therefore need quite large heating radiators, such as for heating, so that they can work at low temperatures.

Ecological payback

To increase temperatures, heat pumps require drive energy (W) far below the amount of heating energy (Q) provided by the system. Q/W expresses the relationship between energy output and input. If that figure is three, then the heat pump provides three times as much heating energy as it consumes itself. While that may seem like a good performance, if the heat pump runs on electricity, we need to keep in mind that only a third of the primary energy used to generate electricity actually reaches your wall socket (see 1.9). In other words, if the energy payback of a heat pump is around three or four, the heat pump merely makes up for the energy lost in power generation.

If the pumps are to perform better than a condensation boiler, then the electric heat pump would have to have a payback exceeding four – which almost never happens in practice.[2] Heating systems with a heat pump therefore generally perform even worse in terms of carbon emissions than a good gas-fired heating system, though heat pumps may perform better than oil heaters.[3]

Heat pumps thus require ideal conditions to produce acceptable emission values – for instance, buildings with very low energy consumption, large-surface heating radiators, and a heat source with constantly high temperatures.

Electric heat pump with downhole exchanger: does not offset more carbon emissions than a natural gas heater.

Figure 8.1 Heat pumps

Source: Energueagentur NRW

8.2 Solar hydrogen

Hydrogen is not a primary source of energy that could round off the spectrum of renewables; rather, it is a secondary energy carrier that first has to be created from another source of energy. If electricity is used to split water into hydrogen and oxygen by means of electrolysis, more energy has to be used than can be gained when hydrogen is combusted.[4]

Hydrogen can be stored and transported over large distances if necessary. It can also provide energy in numerous ways. In fuel cells, power and heat can be provided very efficiently; (catalytic) combustion[5] provides direct heat; and in particular, hydrogen is a primary candidate for environmentally friendly fuels for vehicles. Two advantages offered by hydrogen are clear: there is no lack of water; and the only exhaust is water (vapour). There are some technical risks (of explosion), but they are not worse than with natural gas.[6] In light of these advantages, scenarios were drawn up for a solar hydrogen economy in the 1980s (see Figure 8.2). Gigantic solar farms (PV or solar thermal) would produce power in sunny regions to create hydrogen by means of electrolysis. The hydrogen would then be shipped elsewhere to provide an environmentally friendly substitute for fossil energy. Back then, 'green hydrogen' was touted as the 'energy of the future'.[7]

Some two decades later, hydrogen remains the energy of the future. Some 35 per cent (transport via pipeline) to 50 per cent (liquid hydrogen by ship) of the energy would be lost if solar power is converted into hydrogen and transported from northern Africa to northern Europe. The price of solar power would effectively rise by a factor of two or three in the process. The chance that solar hydrogen will move out of niche markets and become the dominant energy carrier therefore remains very low.

Hydrogen is not needed to transport solar, wind or hydropower across large distances; modern high-voltage lines are less expensive.

This decade, there will also be no need to use hydrogen to store excess solar or wind power to bridge periods of low production. Despite the fast growth of the solar sector, there are only a few hours in a year with accessible solar and wind power. In the long term, it might be better to tailor power demand to power production (by offering rates that fluctuate according to supply and demand, a process known as demand management[8]) in order to reduce the amount of power that needs to be stored in the first place.

Overall, Nitsch et al estimate that there will be no need for hydrogen to store energy before 2020, if we pursue proper climate policy. If we have consistently pursued a path towards a solar economy, we may have to store some electricity by 2050, with estimates ranging from 5 per cent to 38 per cent of overall final energy depending on the specific assumptions.[9]

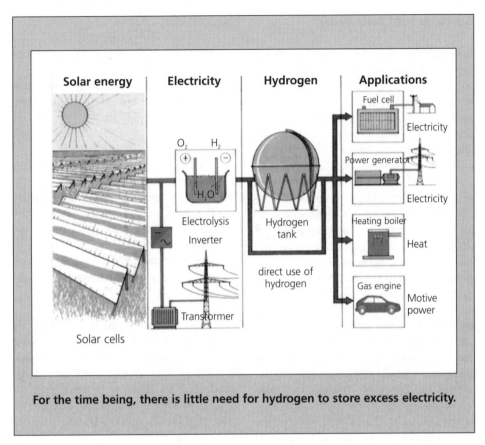

For the time being, there is little need for hydrogen to store excess electricity.

Figure 8.2 Solar hydrogen

Source: Weber, R. *Wasserstoff – Wie aus Ideen Chancen warden*, IZE Reportagen, 1988

8.3 How fuel cells work

Fuel cells have been hot topics in discussions about renewable energy. But fuel cells are not a source of renewable energy themselves; rather, they are an efficient way of generating electricity (and heat) with low emissions. The wide variety of different fuel cells makes them especially interesting for solar energy concepts.

All of the different types of fuel cells[10] have one thing in common: they generate electricity from hydrogen and oxygen. The oxidizing flame that most people are familiar with from chemistry lessons basically occurs under controlled conditions in fuel cells. Hydrogen (or a gas such as natural gas and methanol that contains hydrogen) is input on one side; oxygen (or ambient air), on the other. At the anode, the catalyst splits hydrogen into protons, which pass through the electrolyte, a special membrane that only the protons can penetrate. The electrons flow out of the fuel cell to the electric appliance before passing to the cathode, where they recombine with the protons and oxygen to form water.

In this way, the fuel cell directly generates electric current and heat along with water as a waste product. An indirect process is used in current central power plants (combustion/steam/turbine/generator). Fuel cells are therefore best thought of as a kind of battery that is constantly being 'recharged' with water and oxygen. Because a single fuel cell only has a voltage of around 0.7 volts, multiple cells are stacked together in practice.

Fuel cells offer a number of crucial benefits:

- Their electric efficiency is very high, especially under partial load.
- They are extremely clean. If hydrogen is used, the only byproduct is pure water, but even if hydrocarbons are used, no pollutants (such as sulfur dioxide) are emitted.
- They do not have any moving parts and are therefore quiet.
- Different types of fuel cells allow different fuels to be used – from pure hydrogen to natural gas, methanol, biogas and testified coal.
- Their modular design allows them to be built according to specific power requirements.

Nonetheless, even in 2010, fuel cells still suffer from some technical problems and high costs; they are still being further researched and undergoing field tests. Serial production continues to be postponed, so this technology should not be expected to play an important role any time soon.

Fuel cells can be used in many ways. Stationary applications (cogeneration units), mobile applications in the transport sector ('hydrogen cars'), and small fuel cells for portable appliances (notebook power supplies, etc.) are some of the most common examples. The first products for such applications are already on sale.[11]

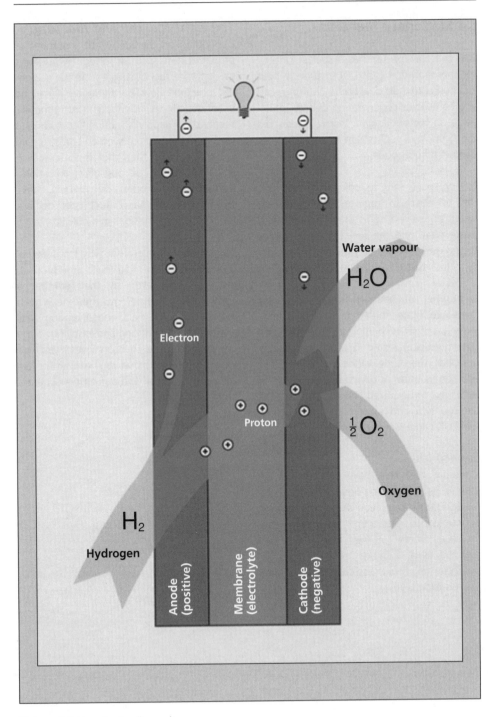

Figure 8.3 How fuel cells work

Source: The authors

8.4 Stationary fuel cells

Fuel cells can be used as a special kind of cogeneration unit (see 1.9) to provide heat and power to individual homes, but they can also be used as cogeneration units in industry or on the public grid. Below, we discuss the example of fuel cells in individual homes and neighbourhoods.

The prospects initially look good: fuel cells can be used to replace classic heating systems. The house is then heated with waste heat, and the fuel cell also supplies the house with efficiently generated electricity, with excess power being sold to the grid.

A number of heating firms and utility companies[12] are therefore working intensively on this technology. Prototypes are currently being tested in practice, but it is not clear when they will hit the market.[13] In addition to issues of cost, a number of problems have to be solved first – such as service life, fuel type and supply structure – before fuel cells can become common.

Service life
Currently, fuel cells have service lives of around 5000 operating hours; the goal is to reach 40,000 hours, equivalent to five years of constant operation. Technically, this increase could come from stronger membranes and thicker catalyst coatings, but the latter would raise costs because platinum is expensive.

The right fuel cell and fuel supply
Depending on the fuel used, a number of problems remain to be solved. Hydrogen is certainly the fuel of choice from the ecological point of view. Furthermore, it can be used directly in fuel cells. Unfortunately, there is no supply structure for hydrogen yet,[14] and the costs of such an infrastructure would be high. Methanol is toxic, which limits the range of possible applications. Furthermore, model calculations have shown that methanol fuel cells do not reduce carbon emissions in a holistic view.[15]

The question is therefore why large power providers have spent so much time promoting fuel cells even as they combat a comparable technology that has been market-ready for years – cogeneration units – with predatory pricing and lobbying. If you are able to make such an investment, we would recommend that you not wait for fuel cells, but go ahead and purchase a cogeneration unit today.

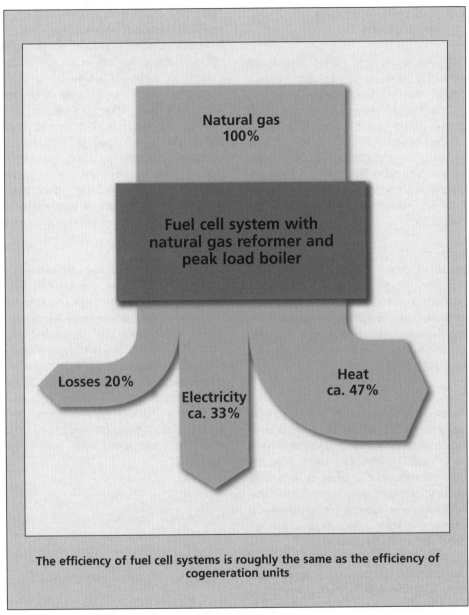

The efficiency of fuel cell systems is roughly the same as the efficiency of cogeneration units

Figure 8.4 The efficiency of fuel cells for domestic power

Source: Authors' depiction based on *Deutscher Bundestag*, 14/5054, p86

8.5 Fuel cells in mobile applications

'The only thing that comes out of the exhaust is water vapour!' In light of climate change and smog, such statements seem attractive at first. No wonder most major car manufacturers have been working on fuel cell cars at some point or another;[16] these cars would have electric drive trains powered by electricity from fuel cells. A number of different concepts have been investigated: fuel cells with hydrogen, methanol and natural gas.

In the mid-1990s, the first prototypes were presented. Daimler-Chrysler also launched a test fleet of 30 buses and 60 cars in 2002/2003.[17] Volkswagen has discontinued its research on hydrogen drive systems and BMW discontinued its field tests on hydrogen as a fuel for cars with conventional engines in 2009 .[18]

Today, no market launch is in sight.[19] One reason is the high cost; fuel cells for cars still cost around 100 times as much as an internal combustion engine. In addition, there is no hydrogen infrastructure, and the cost has been estimated at €80 billion for a network of 2000 hydrogen filling stations – and that would only cover densely populated areas in Germany. In light of such figures, fuel cell hype has died down considerably.[20]

Not even the environmental benefits are convincing if we look at them closely. While fuel cell cars are very efficient, especially in partial load, a lot of energy is needed to provide the fuel. When we compare fuel cell cars running on various types of fuel to conventional cars in terms of environmental impact, the benefits are slight, and technical progress with conventional cars is also expected in the years to come.[21]

One crucial aspect is often overlooked when assessing hydrogen-powered fuel cells. If renewable electricity is used to generate hydrogen for a fuel cell in a car, the car makes do with less conventional fuel, leading to 190g fewer emissions of climate gases per kilowatt-hour.[22] But because that kilowatt-hour was not sold to the grid, other power plants will have to generate it, increasing emissions by 590g (see Figure 8.5). In a holistic view, the use of 'green hydrogen' does not actually reduce emissions in our current power supply system, but rather increases emissions of climate gases.[23]

If at all, hydrogen produced from renewable power for fuel cell cars is a very long-term option for the Solar Age. The highly detailed publications about research into and tests of emission-free fuel cell cars[24] therefore should not mislead us into believing that the acute problems in our transport system can be solved this way. Car travel needs to be avoided more often, conventional cars need to become more efficient and emit less pollution, public transport needs to be used more often, and speed limits need to be reduced.

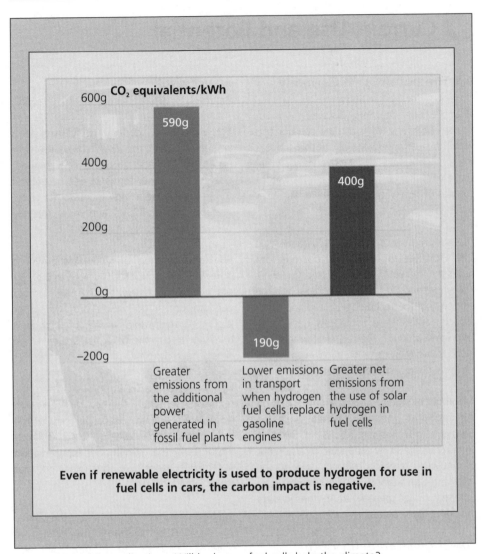

Figure 8.5 Mobile applications: Will hydrogen fuel cells help the climate?

Source: Authors' depiction based on Wuppertal Institut für Klima, Umwelt, Energie GmbH

9 Current Use and Potential

9.1 The potential in Germany

The previous chapters discussed various sources of solar energy and the components within a solar energy supply. In this chapter, we provide a more holistic overview – what share of our energy supply can the various sources of renewable energy have, and how much energy can they provide?

The supply of solar energy is gigantic (see 1.6). However, the theoretical potential can never be completely exploited because availability depends on time and location, the efficiency of the technologies used and other aspects. After these limitations have been taken into account, the technical potential is much lower than the theoretical even before we have discussed economic feasibility.

A number of studies have been produced on the technical potential of renewables in Germany, with varying results. In 2004, a comprehensive study conducted by a number of renowned research institutes came to the following conclusions[1] (all figures per year; potential under greater nature conservation requirements in parentheses):

Power generation (in TWh)

Hydropower	25	(24)
Onshore wind	65	(45–55)
Offshore wind	110	(110)
Photovoltaics	105	(105)
Biomass	85	(70)
Geothermal[2]	66–290	(66–290)

Heat (in PJ)

Biomass	525	(425)
Solar thermal	1040	(1040)
Geothermal	1175	(1175)

Fuels (in PJ)

Biomass	490	(320)

This potential is currently only used to a limited extent. Of the 25TWh of hydropower potential, 21 have already been exploited, but the only other forms of energy utilized to a great extent are biomass for heat (around 300 of 525PJ) and more recently onshore wind power (40 of 65TWh).[3]

The total potential of renewable power is 350–719TWh, roughly equivalent to 62–127 per cent of current German power consumption (2003: 585TWh). In other words, a renewable supply of power is hardly a utopia; the potential is there – we simply need to use it.

Depending on the type of biomass used, the renewable potential for heat lies between 42 per cent and around 60 per cent of the fuel consumption level of 2003 (roughly 5300PJ). If the tremendous conservation potential in space heating is exploited, that share would increase.

In terms of current consumption, the potential for fuels is the lowest. But even if all biomass is dedicated to fuel production, we would only be able to cover a third of our current consumption in the best case. While we can import biofuels, we will not be able to achieve a renewable fuel supply unless we accept less mobility and greater vehicle efficiency (see 5.10).

Overall, a holistic view reveals that a solar strategy can only be successful if we greatly increase energy efficiency and conservation (see 1.8).

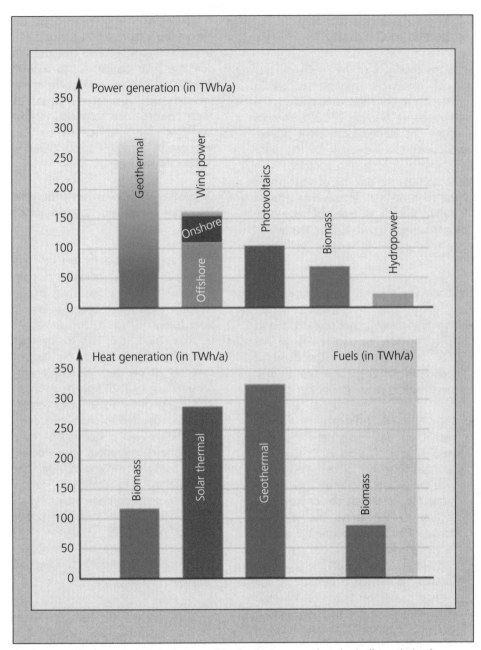

Figure 9.1 Technical potential of renewables in Germany and ecologically optimized scenario

Source: DLR, ifeu, Wuppertal Institute, 2004

9.2 The future has already begun in Germany

Since the beginning of the 1990s, production of renewable energy has grown considerably in Germany. Wind power is the most impressive example. By 2005, 14 years after the first feed-in tariffs were offered, wind power had increased 100-fold since 1992. Photovoltaics is also booming. From 1990 to 2005, the amount of solar power generated in Germany increased by a factor of 1000.[4] But other types of renewable energy have also grown at rates that would make other industries envious (see Figure 9.2).

This boom was the result of specific policies:

- The Power Feed-in Act of 1991 and its successor, the Renewable Energy Act (EEG, see 11.9) ensure that producers of renewable power get a return on their investment in renewable power generation.
- The 100,000 Roofs Programme provided crucial upfront financing for photovoltaics from 1999 to 2003. Germany now has companies all over the country with the expertise to install solar power arrays. In addition, the cost of the manufacture of PV arrays has dropped considerably, thereby paving the way for further growth.
- Part of the proceeds from the Ecological Taxation Reform were devoted to the German government's Market Incentive Programme, which had an annual budget of around €200 million (2005). In addition to solar thermal systems, energy from biomass was also supported.

- A growing number of tradespeople are discovering solar as a business field and working to convince their customers to purchase solar equipment. To support them in this endeavour, Germany implemented such national campaigns as 'solar power – now's the hour', 'solar heat plus'[5] and 'heat from the sun', an information campaign across Germany about solar thermal.

For a number of reasons, the renewable energy sector will continue to grow quickly in the next few years:

- New production plants for solar cells and solar panels have increased supply. Production capacity for solar power technology has grown more than tenfold since 1999.[6]
- In June 2009, the EU's Renewables Directive took effect with the goal of covering 20 per cent of final energy consumption by 2020 from renewables.
- At the level of the EU, all utility companies are obligated to show consumers how their power is generated.[7] As a result, the debate about where power comes from has been revived. A number of households and businesses therefore decide to purchase clean power, which would help increase renewable power production (see 12.2).
- Solar thermal systems are now of very high quality according to German consumer protectionists.[8]

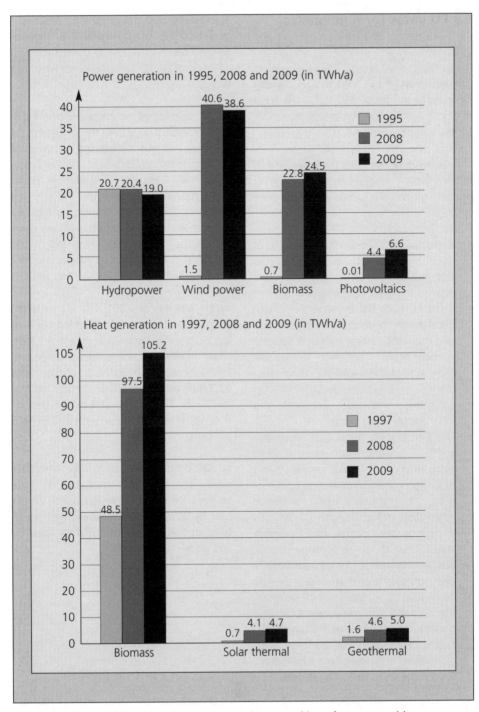

Figure 9.2 Germany, the transition has begun: Power and heat from renewables

Source: BMU, Erneuerbare Energien in Zahlen, 2009

9.3 EU votes for renewables

In its White Paper for Strategy and Action Plan for renewables,[9] the European Commission stated in 2001 that renewables have not been sufficiently used in the EU; it therefore calls for urgent action in policies for the European Union.

The Commission believes there are several reasons why a comprehensive strategy to promote renewables can no longer be done without:

- Currently, the EU imports 50 per cent of its energy. If nothing is done, the Commission estimates that this share will rise to 70 per cent by 2020.
- If the EU does not manage to increase the share of renewable energy considerably (from the current level of 6 per cent to more than 12 per cent), the EU will have a hard time fulfilling its environmental protection obligations in compliance with both European and international agreements.
- The Commission holds that renewable energy is an important future market. Since other countries, such as the US and Japan, are currently implementing measures to support their domestic renewables industries, the EU believes there is a great danger that European industry might lose its leadership position in this sector. 'Unless we have a clear, comprehensive strategy followed up by legislation, there will be delays in the growth of renewables.'[10] The Commission also feels that a 'long-term, reliable framework for the growth of renewables' is an important requirement for companies to make investments.

Against this background, the EU adopted a goal in 2001 of increasing the share of renewables in total energy consumption to 12 per cent by 2010. As the White Paper put it:

Considering all the important benefits of renewables on employment, fuel import reduction and increased security of supply, export, local and regional development, etc. as well as the major environmental benefits, it can be concluded that the Community Strategy and Action Plan for renewable energy sources as they are presented in this White Paper are of major importance for the Union as we enter the 21st century.[11]

In the autumn of 2001, the EU therefore adopted its Directive on the Promotion of Electricity from Renewable Energy Sources, which states that the share of renewable electricity within the EU is to increase from 13.9 per cent in 1997 to 22 per cent in 2010.[12] By 2007, renewable power generation within the EU-15 had increased to 16.6 per cent.

In 2009, the EU adopted more ambitious goals. The share of renewables in total energy consumption is to increase to 20 per cent by 2020. Furthermore, biofuels are to cover at least 10 per cent of total gasoline and diesel consumption.[13]

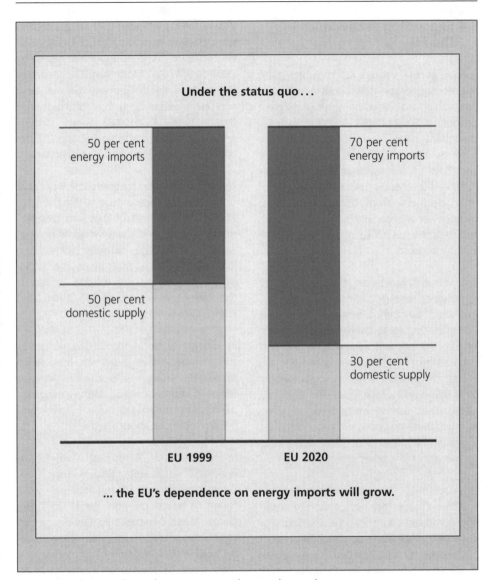

Figure 9.3 The EU's dependence on energy imports is growing

Source: Kommission der Europäischen Gemeinschaften, Grünbuch, 2001

9.4 Expanding renewables in the EU

A look at the various EU member states reveals considerable differences in the use of renewables, mainly as the result of different geography. For example, countries with tall mountains have much more hydropower potential than flat countries. It therefore comes as no surprise that Austria (65 per cent) and Sweden (44 per cent) get a lot of their electricity from hydropower, while hydropower is negligible in The Netherlands and Denmark, which each have only 0.1 per cent hydropower.[14]

But the differences in the potential of renewable energy cannot simply be explained by the differences from one European country to the other. For instance, France, the UK and Ireland have the best conditions for wind power. A wind turbine in Ireland can produce twice as much electricity as one in an average location in Germany. Nonetheless, Germany produced four times as much wind power as Ireland and the UK together in 2008, and installed wind capacity was about five times greater.[15]

Unsurprisingly, solar thermal is used in southern Europe more often than in northern European countries like Sweden. For instance, Greece made up about 14 per cent of the collector area installed in EU in 2008. It thereby only came in third, however, behind Germany, which made up about 40 per cent of the collector area installed in Europe with 11.3 million square metres. In terms of per capita installed collector area, Austria came in first with a bit more collector area than Greece. In contrast, sunny countries such as Italy and Spain do not even have 5 per cent of the solar thermal market.[16]

Within the EU, photovoltaics has boomed in the past few years. In 2004, roughly 1GW was installed, but that figure had grown by 2008 to 9.47GW. More than 90 per cent of the systems are in Germany (56 per cent) and Spain (36 per cent), both of which have feed-in tariffs. In contrast, sunny Portugal and Greece only made up 0.7 and 0.2 per cent of the European market, respectively.

Overall, renewables covered some 9 per cent of final energy consumption within the EU in 2006. Just over half of that was biomass, primarily wood and waste wood. In Finland, Sweden and Latvia, biomass makes up a large part of the national primary energy pie. But in terms of absolute volume, biomass is the largest in Germany (151TWh) and France (135TWh) in 2007.[17]

In addition to natural conditions, policies in the member states determine how much renewable energy is produced. Another factor is domestic fossil energy resources, decisions for or against nuclear power, and general attitudes about energy.[18]

In 2004, the European Commission reviewed renewables targets for 2010. Denmark, Germany, Finland and Spain were found to be on the right path.[19] In 2007, Germany and Denmark had already reached their targets for 2010 for renewable electricity, while France and Italy had not increased their share of renewable electricity at all from 1997 to 2007 (See Figure 9.4).

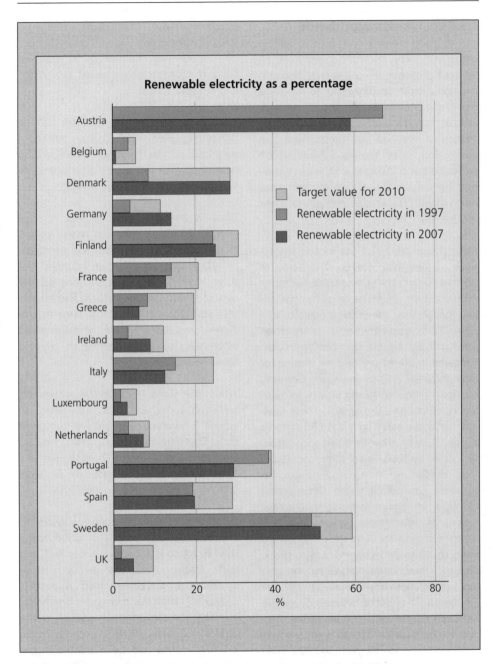

Figure 9.4 Expanding renewables in Europe: Power generation in EU-15

Source: BMU: Erneuerbare Energien in Zahlen, 2009

9.5 Renewables worldwide

Renewable energy currently covers some 17 per cent of global primary energy demand. Traditional biomass (firewood for cooking and eating) makes up the largest piece of that pie; unfortunately, such practices cause irreversible damage to forests, and the smoke from open fires is a health risk.[20] Researchers and industry are therefore now working on second-generation biomass, which is both sustainable and environmentally friendly.

Water power makes up the second biggest block of renewable energy. Worldwide, it provides 90 per cent of renewable electricity. Wind and solar power are small by comparison, though they are growing quickly. From 2000–2008, grid-connected solar arrays grew globally by 60 per cent per year, compared to 20–30 per cent per annum for wind turbines. Policies in Japan, Germany and Spain produced strong growth for grid-connected photovoltaic arrays. These three countries alone made up around 80 per cent of the 13GW of grid-connected solar capacity installed by 2008. But photovoltaics is also used outside of the Organisation for Economic Co-operation and Development countries. In regions without grid access, millions of Solar Home Systems (see 3.3) provide electricity for domestic consumption. Many countries (India, China, Egypt, Brazil, Peru, etc.) have comprehensive support policies. In Bangladesh alone, a project will install 1.3 million Solar Home Systems. There has also been tremendous progress in the planning and construction of solar thermal power plants over the past few years. In 2008, 8GW was either in planning or under construction, 6GW of which was in the US alone.[21]

Micro-hydropower, wind power and biomass cover roughly the same amount of power production. China comes in first with micro-hydropower; indeed, the Chinese have more than half of global installed capacity, often to supply power to a village without grid access.

In 2008, the US, Germany and Spain had the greatest installed wind capacity, but countries such as India, Italy and China have also now discovered this source of environmentally friendly power (see Figure 9.5).

China is also the leader in solar thermal energy. Simple systems are manufactured locally at low prices, making them affordable already without any further subsidies. Not surprisingly, China has more than 65 per cent of global installed capacity.[22] Israel also uses solar thermal arrays to a large extent; indeed, it is the only country in the world where solar hot water systems are required for all new buildings.[23]

Worldwide, some 79 billion litres of biofuels were produced in 2008, equivalent to around 3 per cent of global fuel consumption. The leading ethanol producers are Brazil and the US (see 5.9), with Germany and the US being the leader for biodiesel (around 12 billion litres worldwide).

Efforts are being made around the world to expand renewable energy. In 2008, some US$120 billion was invested in this field, and that figure does not include large hydropower, which entailed a further US$40–45 billion. Far more was nonetheless invested in conventional energy: some US$150 billion (in 2004). Worldwide, some 2.4 million jobs had been created in the renewable energy sector by 2006, but this figure has increased dramatically since though reliable global figures are not yet available.[24]

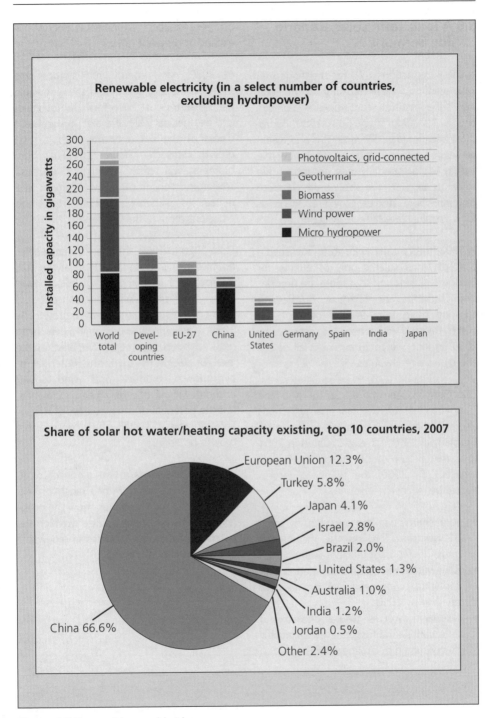

Figure 9.5 Renewables worldwide

Source: REN21: Renewables Global Status Report 2009

9.6 A long-term solar scenario for Germany

In the Introduction (1.7), we mentioned an early scenario for the transition to the Solar Age –the Institute for Applied Ecology's Energy Transition study from 1980. Today, three decades later, the strengths and weaknesses of various sources of renewable energy – and especially their potential – are easier to describe. We also now have more comprehensive studies providing scenarios for a sustainable solar energy economy.

In 1998, the Wuppertal Institute published a concept for a solar supply of energy for Europe.[25] They found that there are no principal technical or financial hurdles to covering more than 90 per cent of our energy demand with renewable resources by 2050. In fact, we could even get all of our energy from renewables.

The German Aerospace Centre and the Fraunhöfer Institute for Solar Energy Systems also published a long-term solar scenario for Germany in 1997.[26] It was updated in 2004 to take better account of ecological aspects.[27] By 2050, the following targets should be reached:

- Total energy demand is to drop by some 51 per cent through the incremental implementation of efficient technology.[28]
- Renewables are to be expanded significantly to cover nearly 45 per cent of demand by 2050.
- Nuclear power is to be phased out before 2030 and fossil energy will mainly be consumed in cogeneration units.

Different sources of renewable energy will expand at different rates.

By 2020, wind energy and biomass will make up the largest share of the renewable pie. Solar thermal collectors, photovoltaics and geothermal will not make up significant pieces of the pie until then despite fast growth rates (see Figure 9.6). Fuel from biomass will grow prudently because stationary consumption of biomass will remain less expensive and reduce carbon emissions more in the beginning.

From 2020 to 2050, the share of renewables doubles in this scenario, with solar thermal, photovoltaics and imports of solar power (either from photovoltaics or concentrated solar power) posting the greatest growth. By 2050, carbon emissions will have been reduced by 80 per cent compared to the level of 1990. By then, the potential of hydropower, biomass and wind power onshore will have largely been exhausted, but the potential of other sources of renewable energy will still offer considerable growth potential.

Of course, any scenario with a horizon of 40 years must be taken with a pinch of salt. Nonetheless, such scenarios show the possibilities, costs and time frames in which the potential of individual sources of renewable energy can likely be tapped.

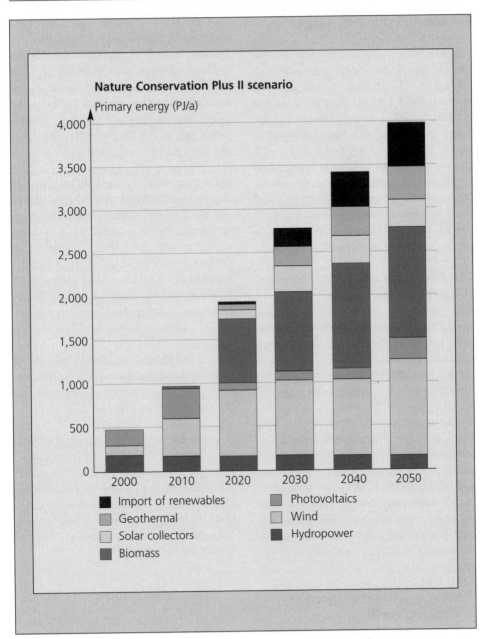

Figure 9.6 A long-term scenario for Germany: Renewables as a part of primary energy consumption

Source: DLR, ifeu, Wuppertal Institute, 2004

9.7 The 100 per cent target

As previous sections have demonstrated, we can get all of the energy we need from renewables. Nonetheless, it will still take several decades for this target to be reached in many areas. It is therefore crucial for further progress towards the Solar Age that we not lose sight of the target: completely phasing out nuclear and fossil energy. One important first step is for individual communities, regions and countries to attain a 100 per cent renewable energy supply – or work towards one for the foreseeable future.[29] We already have some encouraging examples.

The bioenergy village of Jühnde – 100 per cent biomass

The village of Jühnde near Göttingen, Germany, has been providing its 850 inhabitants with power and heat using almost exclusively renewable energy since 2005. A community-owned biogas unit (700kW) and a woodchip-fired heating plant (550kW) provide the energy. Heat is distributed through a network of pipes with a total length of 5.5km. These units are fired exclusively with local resources, which boosts the local economy. Biogas comes from 800 cows and 1400 pigs on nearby farms, while grass, garden waste and dedicated energy crops (rapeseed, corn and sunflowers) are also used. The village's cogeneration unit generates some 4 million kilowatt-hours of power each year.

Güssing – renewables provide economic growth

Güssing is a town in eastern Austria. Local companies were taking advantage of the close border to Hungary, with the result that local jobs were being lost. In 1988, the Burgenland region had become the poorest region of Austria. In 1990, the community council of Güssing therefore resolved to get all of its energy from renewables.

Over the years, buildings have been renovated to reduce energy consumption, a biodiesel unit was installed, two district heating networks have been set up, and Austria's largest wood-fired biomass unit went into operation. Because the town had piped heat to spare, it attracted 50 new businesses, which created more than 1000 new local jobs.[30]

In the past few years, some 90 communities and regions in Germany have set a goal of 100 per cent renewables. They are working to improve regional added value, increase energy security and protect the climate. A research project at the University of Kassel is studying which factors helped Germany get to 10 per cent and how that knowledge can be shared to bring the country to 100 per cent.[31]

Iceland already gets around 75 per cent of its primary energy consumption from renewables. The island has more hydropower and geothermal energy than it can use.

To reach the 100 per cent target, the transport sector will have to be restructured. Hydrogen will have to be made from renewable electricity as a fuel for vehicles. The first pilot projects with cars and buses[32] and a ship[33] have already started. Unfortunately, the economic crisis that started in 2008 has slowed down progress.

9

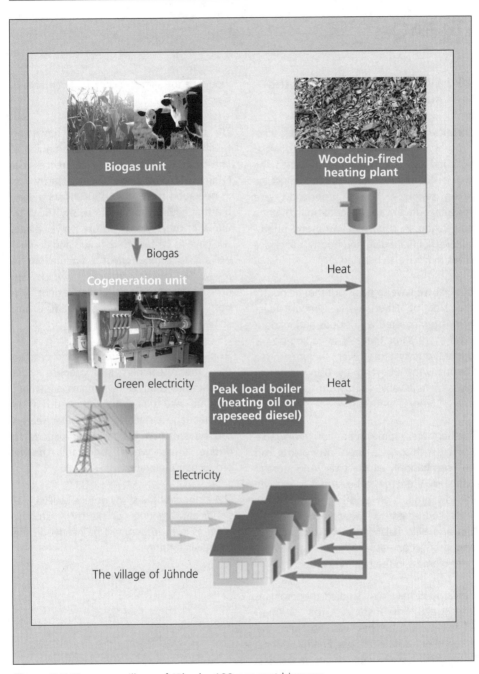

Figure 9.7 Bioenergy village of Jühnde: 100 per cent biomass

Source: Authors' depiction based on IZNE, Bioenergiedörfer, Göttingen, 2006

10 FAQs

10.1 What do we do when the sun isn't shining?

Sceptics often point out that renewable electricity is intermittent. The question then is how industry is to manufacture when the sun is not shining. This problem must be taken seriously. Power production and consumption always have to match on a power grid. And because wind and photovoltaics fluctuate daily and hourly, they leave holes that have to be filled.

But first we have to point out that a country the size of Germany is almost never completely covered with clouds and without wind at the same time. By spreading power generation systems over a large area (Germany/the EU) and by using different types of renewable energy, the problem of intermittency is reduced.

Furthermore, some types of renewable energy, such as large run-of-river plants, can be used basically all the time. And storage dams allow energy to be stored to provide for low production at other times. Biogas units connected to cogeneration systems can also help balance power consumption and production, as can power generated from biomass in the heating plants.

Researchers have also studied this problem intensively. For instance, the German Aerospace Centre has worked with the Fraunhöfer Institute for Solar Energy Systems to see how power demand can be covered in a system of mainly renewable energy sources.[1] The researchers assumed that 25 per cent of power would come from cogeneration in 2050, with 60 per cent coming from renewables. They also assume that this would be the same quality of power that we have today.

The researchers found that the remaining fossil fuel power plants would be running at far lower capacity. In the reference year (1994), 55GW of power plant capacity was in operation more than 4500 hours a year, but that figure would fall to below 15GW by 2050. In other words, future power plants will have to be easy to ramp up and down if we are to compensate for fluctuations in renewable power production, which can only be controlled to a limited extent. The best option here is gas turbines and combined cycle plants.[2]

To match power consumption and production, the researchers also propose more advanced methods of demand management than are used today.[3] Appliances such as air conditioners, refrigerators and water heaters can be switched off for brief periods to bridge times when renewable power production is dipping.

The authors of the study do not believe that significant amounts of electricity storage (such as in hydrogen) will be needed in the foreseeable future.

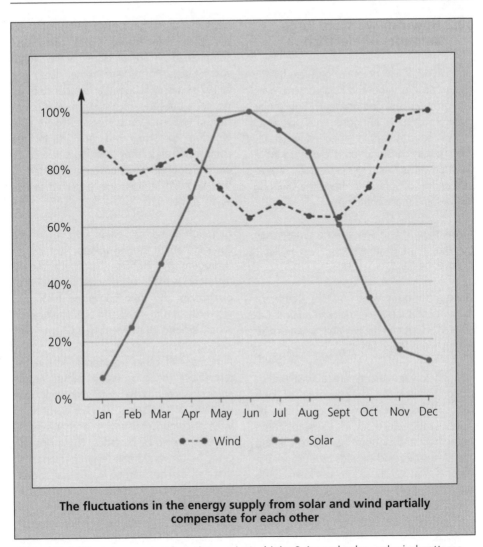

The fluctuations in the energy supply from solar and wind partially compensate for each other

Figure 10.1 What do you do when the sun isn't shining? Annual solar and wind patterns

Source: Kohler, Leuchtner, Müschen, *Sonnenenergiewirtschaft*, 1987

10

10.2 How can we store large amounts of electricity?

Up to 2007, it was no major problem for the grid to absorb wind and solar power, even though they are intermittent. But today, there are times when more renewable electricity is generated than is needed, especially when power consumption is low and a lot of wind power is being generated. As renewable energy continues to grow, the situation will become more common.

Furthermore, there will also be times when not enough renewable electricity is available to cover demand.

There are various ways to store significant amounts of excess electricity in such situations so that it can be available when power production is lower later.[4]

Pumped storage plants are a well understood technology. They consist of two basins, one above the other, connected by pipes. Excess electricity is used to pump water from the bottom basin into the top one. When demand exceeds production, the water is then let down into the lower basin again; on its way, it drives hydropower turbines. Germany already has more than 30 such power plants with an efficiency of up to 85 per cent and an overall capacity of 6GW. The potential for these power plants is, however, already largely exhausted in Germany, though there is potential for expansion in Switzerland and Scandinavia.

Compressed air energy storage (CAES) is less well known. In this technology, excess electricity is used to compress air up to 100 bar in subterranean salt caverns. Wind power is needed, the compressed air essentially serves the same function as the compressor stage of the gas turbine, which reduces the turbine's natural gas consumption by 40–60 per cent (see Figure 10.2).[5] The first compressed air power plant (in the world) was set up just outside of Bremen, Germany, in 1978.[6] It only achieved an efficiency of 42 per cent, but modern systems have efficiencies up to 55 per cent. Researchers are now working to store the heat from the compressed air so they can feed it back to be out flowing air wind power as needed. If they succeed, efficiency is expected to rise up to 70 per cent.

Although there has been a lot of media coverage about storing excess electricity in hydrogen, this option is not effective. If hydrogen is made from water by means of electrolysis and then converted back into electricity in a fuel cell, the overall efficiency is only around 25 per cent (see Chapter 8).[7]

A number of other approaches still in the test phase focus on using existing appliances/applications to store electricity. One example[8] is hybrid cars, which have both an internal combustion engine and an electric motor powered by batteries. These cars can easily be converted into 'plug-in hybrids' so that they can be charged from a wall socket. Excess electricity could then be stored in thousands of car batteries, but the cars could also provide electricity when need be. If a large number of such relatively small batteries work together, we end up with a way to store a large amount of electricity. For instance, 100,000 plug-in hybrids could store or generate around 1000MW of electricity.

Large refrigerated warehouses can also be used to store electricity. When there is excess electricity, they can cool down a bit more, and when less renewable electricity is available they simply stay off longer.

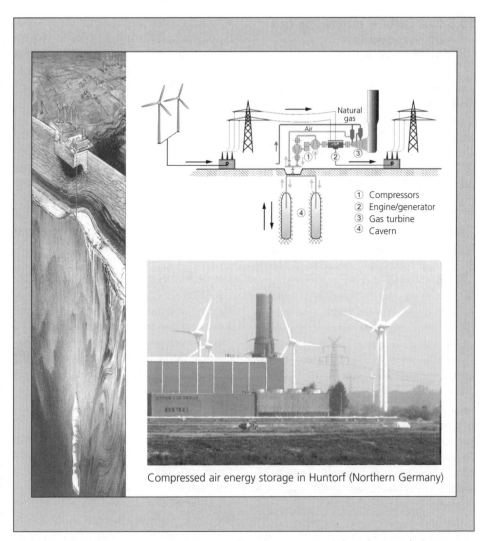

Compressed air energy storage in Huntorf (Northern Germany)

Figure 10.2 How much energy can be stored? Energy storage with compressed air systems

Source: Fritz Crotogino, KBB Underground Technologies GmbH, Hannover

10.3 Can carbon emissions not be avoided less expensively?

Every time fossil energy is burned, carbon dioxide (CO_2) is emitted. Along with other greenhouse gases, CO_2 causes the greenhouse effect, which is changing our climate and increasing the average temperature on the Earth. There are basically four ways to prevent carbon emissions.[9]

1 Making fuel consumption more efficient and conserving energy.
2 Using fuels with less carbon content (such as natural gas instead of oil and oil or gas instead of coal).
3 Increasing the efficiency of energy conversion (for example, by making power plants more efficient and using cogeneration instead of getting electricity from conventional condensation plants).
4 Using more renewable energy.

Each of these actions entails costs, on the one hand, and a number of benefits (in addition to lower carbon emissions) on the other. To correctly determine the carbon avoidance costs, these costs and benefits have to be weighed off against each other – which is not easy.

Furthermore, we have to keep in mind what angle we come at the calculation from. If we have a macroeconomic view, the results will be different than those calculated by private investors or utility firms. Furthermore, the external costs of our energy supply (see 11.3) must be included in the calculation. Although attempts to calculate carbon avoidance costs all have their drawbacks,[10] we will take a look at the general idea below.

The carbon avoidance costs are much higher for photovoltaics than for wind energy, solar thermal or energy conservation. In other words, the cost of avoiding the emission of a ton of carbon is several times greater than for other technologies at present.

Under German feed-in rates, however, the situation looks different for an average German homeowner, who also has easy access to low-interest loans for PV investments. Carbon avoidance costs are therefore very low for such investors. But whatever the case, investors generally do not base their decisions on carbon avoidance costs.

The benefit that investors get from photovoltaic arrays is not limited to the money they get in return for their solar power, but also extends to the feeling of having made a useful investment that is good for the environment, helps foster a young industry, and provides a little bit of independence from utility companies.

Such households may have been able to have the same positive environmental impact with energy-saving technologies at a lower cost, but the fact that they chose a photovoltaic array shows that subjective assessments of costs and benefits are not the same thing as theoretical macroeconomic assessments of costs based purely on economics.

10

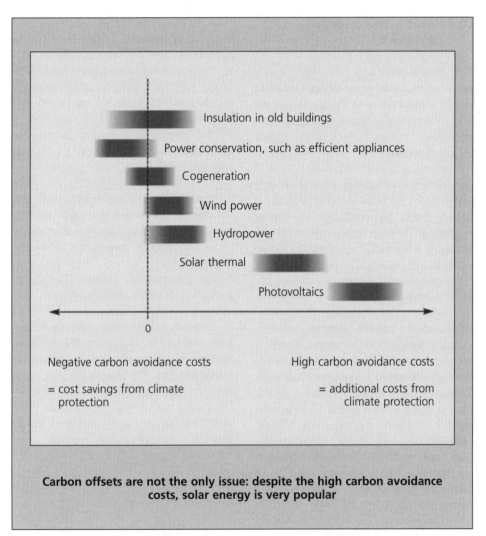

Insulation in old buildings

Power conservation, such as efficient appliances

Cogeneration

Wind power

Hydropower

Solar thermal

Photovoltaics

0

Negative carbon avoidance costs

= cost savings from climate protection

High carbon avoidance costs

= additional costs from climate protection

10

Carbon offsets are not the only issue: despite the high carbon avoidance costs, solar energy is very popular

Figure 10.3 Can CO_2 be offset less expensively in some other way? CO_2 avoidance costs (in euros per ton)

Source: The authors

10.4 What is the energy payback?

With wind turbines and solar arrays getting bigger all the time, some wonder whether all of the materials used ('grey energy') are worthwhile – or, put differently, do we invest more energy in such systems than they will be able to generate?

The term to understand here is 'energy payback'. It is an indication of how long the system needs to produce the energy invested in it. After that time, it begins producing a 'surplus'. If a system can remain in operation longer than it needs to pay back the original energy investment, the energy payback is positive.

Energy payback not only depends on manufacturing, but also on how the system is used. For example, a wind turbine in windy locations will produce more power, thereby foreshortening its energy payback. Pick and Wagner[11] have calculated the energy payback of a wind turbine under different wind velocities and different tower heights. They found that it took the turbine between 3.3 months on the coast (wind velocities of 6.87m per second) and 6.2 months for a turbine inland (65m up, with winds at 5.91m per second). The Institute of Applied Ecology came to similar conclusions.[12]

The energy payback of a solar thermal system depends on the material used, the amount of sunshine and the solar coverage rate – but also on the heating system it partially offsets. If the heating system is old and inefficient, the collectors will pay for themselves within less than six months. But if a condensation boiler is offset, the energy savings are lower and the energy payback is longer. Depending on the material used and the amount of sunlight, energy payback generally ranges from 0.6–2 years.[13]

With the crystalline solar cells generally used today in photovoltaic arrays, the energy payback is generally around three to four years north of the Alps and two years south of the Alps depending on solar conditions and what is included in the calculation. But these figures are falling all the time, and by 2010 systems north of the Alps have probably already reached an energy payback of only two years.[14]

In conclusion, energy payback depends on a number of factors and is hard to state in general. But all of the available data clearly show that systems running on renewable energy produce several times as much energy as was invested in them. Their energy payback is therefore clearly positive.

In contrast, nuclear and coal plants also require a fuel in addition to the materials and energy for construction. Because these plants produce less energy (as electricity) than is contained in the fuels consumed, we are always putting more energy into these plants than we get out of them. Their energy payback is therefore always negative.

10

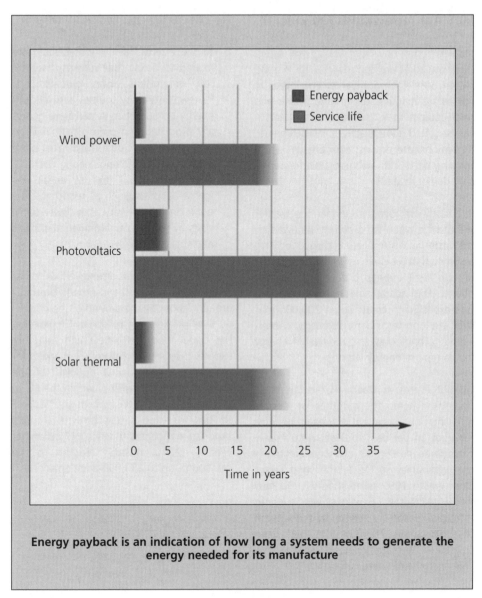

Energy payback is an indication of how long a system needs to generate the energy needed for its manufacture

Figure 10.4 Comparison of service life with energy payback

Source: Institut für Elektrische Energietechnik Berlin, 1996, and Photon, Sept 2005

10.5 Are renewables job killers?

Studies about how renewables will affect the job market are controversial for a good reason. While the number of employees at companies manufacturing wind power and solar equipment – and even at suppliers – can be stated quite exactly, it is much harder to demonstrate downstream effects, which certainly occur, but are impossible to empirically demonstrate.[15]

In a study on how the German job market will change when we switch to renewables, the Institute of Applied Ecology not only assumed that renewables would offset some of our fossil energy, but also all nuclear power. This sustainable energy approach was expected to create some 200,000 jobs over the long term because energy conservation technologies and renewable energy would replace energy imports.[16]

Current events in Denmark confirm these findings. There, the number of people receiving unemployment or welfare benefits fell from 14 per cent to 6 per cent, mainly because nearly 65,000 jobs were created in the renewables sector. In the wind sector alone up to 1999 more than 20,000 new jobs were created.[17] Exports of wind turbines were one reason.[18] Likewise, by the autumn of 2008 some 280,000 jobs had been created in Germany in the fields of renewable energy and energy efficiency.[19]

The jobs created offer two crucial benefits:

1 To the extent that the jobs created are related to a domestic market, they are not dependent upon globalization. Renewable energy sources and efficient energy consumption systems are generally produced and used locally. A large part of the work performed takes place in local trades and engineering firms.

2 The jobs created do not entail any follow-up costs (such as road construction); on the contrary, they lower social costs by reducing environmental impacts and social conflicts.

Another benefit that is often overlooked has to do with innovation and growth potential brought about by the growth of the domestic sales market. The European Commission has understood this aspect and therefore expects the growth of domestic markets for renewable energy and a doubling of the share of renewable energy within the EU to provide European industry with great opportunities for exports and growth. 'Annual exports are expected to reach €17 billion per year by 2010, possibly leading to the creation of up to 350,000 additional jobs.'[20]

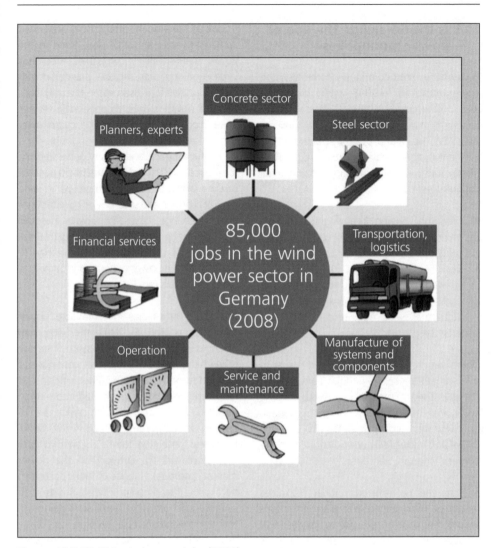

Figure 10.5 85,000 wind power jobs (2008)

Source: Eigene Dartstellung nach: VDMW and BWE: Windenergie in Deutschland

10.6 Is the Solar Age the end of power monopolies?

In Germany, eight companies dominated the power market up to 1998, when the market was liberalized. That year, RWE, PreussenElektra, VEAG, EnBW, Bayernwerk AG, VEW, HEW and BEWAG generated 85 per cent of the electricity in Germany. They mainly had two ways of marginalizing small competitors:

- Predatory pricing kept highly efficient, environmentally friendly cogeneration units off the market.[21]
- The Feed-in Act (11.9), which had levelled the playing field for wind power and hydropower, was considered unconstitutional, and numerous court cases were filed.

Liberalization further concentrated power in a smaller sector – VEBA and VIAG (PreussenElektra) merged in 1999 to become E.ON; RWE and VEW also merged.[22] Overall, only four large conglomerates (RWE, E.ON, EnBW and Vattenfall) currently dominate the power market in Germany.

In our energy transition, local resources would be used and power plant structures would be distributed. Here, more local responsibility – and political input – is needed. To what extent have these requirements been fulfilled?

Current data allow us to draw some conclusions.[23] First, the good news: small firms and private investors currently make up some 90 per cent of all renewable power generators in Germany. On the one hand, that figure clearly demonstrates how committed private investors are to paving the way towards the Solar Age; on the other hand, it also shows how reluctant some power providers remain.

A look at the amount of electricity generated with renewables shows a clear trend in the power market. Utility companies – be they large conglomerates, regional private firms or municipal utilities – used to be the only ones running power plants. And because utilities were the ones operating large hydropower dams, these firms also were responsible for a lot of the renewable electricity generated. As recently as 2000, they still made up 60 per cent of the market. But the strong growth in wind energy, biomass and solar power have brought about a change; nowadays, private investors have taken the lead. When it comes to these three types of renewable power, more than 70 per cent of the power generated is owned by private investors (see Figure 10.6).

The share of utilities in renewable power continues to drop. By 2005, the figure had fallen to 35 per cent. At the same time, the market segment of renewables continues to grow. From 1999–2005, the amount of renewable electricity generated more than doubled, reaching a 10 per cent share of energy consumption in 2005.[24] But taken together, these two trends mean that new market players are at work on the power market, and the influence of monopolies can be expected to drop somewhat. But as soon as large offshore wind farms start going up, this trend could reverse because only large firms with deep pockets can finance such projects. A change in taxation also means that relatively few wind turbines are now being put up as community projects.

The new market segments of electricity from solar, wind and biomass may still only be relatively small players, but they are growing very quickly. At the moment, further changes in energy policies may shift power production further into the control of corporations, especially if Germany's feed-in rates specified in the Renewable Energy Act are fundamentally revised.

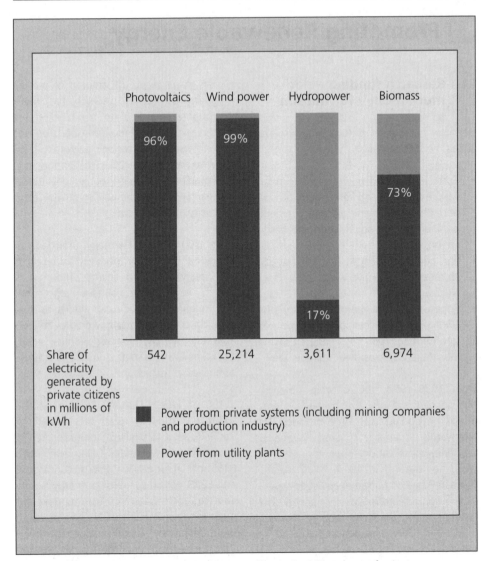

Figure 10.6 The Solar Age: The end of power monopolies? The share of private power generators as of 2004 in Germany

Source: Authors' depiction based on VDW-Projektgruppe Strombilanzen, 2006

11 Promoting Renewable Energy

11.1 Research funding – not much money for the sun

Germany has helped fund energy research across the following fields:[1]

- Fossil energy and power plant technology (extraction and refinement of coal and other fossil energy sources, firing technology, coal liquefaction and gasification, etc.).
- Nuclear technology (reactor safety, storage of radioactive waste, etc.).
- Nuclear fusion.
- Renewables and energy efficiency (the photovoltaics, wind power, biomass, geothermal, fuel cells, energy-saving industrial processes, heat storage, etc.).

From 1956–1998, the German Research Ministry devoted some €23 billion to energy research. Funding was only provided for renewable energy and energy efficiency/conservation after the first oil crisis, i.e. starting in 1974. By then, the German Research Ministry had already spent €2.6 billion researching nuclear energy. The distribution of funds for energy research from 1974–1999 is shown in Figure 11.1.[2] The chart clearly shows the following:

- At nearly €15.3 million, the German government spent nearly five times as much researching nuclear energy than it spent on renewables and energy efficiency/conservation.
- Even after the 1992 UN Conference on the Environment and Development in Rio de Janeiro, which called for sustainable economics, research expenses for nuclear power remained far above the funding for renewables research.

- For many years, the amount of money spent on fusion research has been roughly equivalent to the funding for renewables even though experts do not believe that a fusion reactor will be workable before 2050. Furthermore, it is completely unclear whether any fusion reactor will ever be able to provide electricity at competitive prices.[3]

From 2001–2003, Germany provided an additional €51 million per year for research into renewables and energy efficiency as part of its Future Investment Programme. This research covered such things as fuel cells, drive technologies powered by renewable energy, geothermal, offshore wind, high temperatures solar thermal and energy-optimized construction.[4]

In the 5th Energy Research Programme, the following funding was set aside for 2005–2008: €460 million for fusion; €421 million for renewables (including bioenergy); €455 million for efficient energy conversion; and €219 million for nuclear energy (safety and storage research).[5] The funding for nuclear research has clearly been reduced, but more money was still provided for fusion than for research and development related to renewables.

Within the EU, research funding still clearly favours nuclear energy. In the 6th Research Framework Programme from 2003–2006, two thirds of the budget went to nuclear research.[6] In the 7th Research Framework Programme from 2007–2011 €2.7 billion is dedicated to nuclear research.[7]

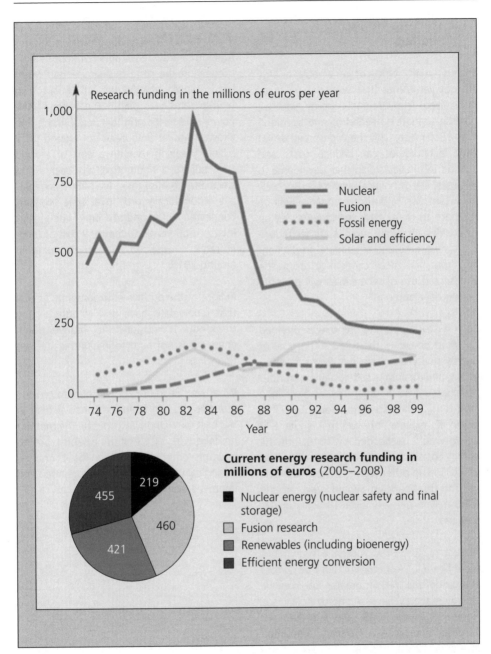

Figure 11.1 Federal funding for energy research in Germany

Source: Authors' depiction based on Energieforschungsprogramm der Bundesregierung, Berlin, 2005

11.2 Start-up financing is needed

Because many types of renewable energy are not yet competitive with the prices of (subsidized) conventional energy sources, political support is needed for the transition to the Solar Age. Only the widespread use of new technologies can reduce costs and provide incentives for further development. The reliability of political support is especially important, for without it companies will be reluctant to make new investments, set up new sales channels, train staff to reduce installation time and streamline production. In return, practical experience from the widespread use of renewables will promote further development.

Such a self-supporting trend was attained for wind power in the 1990s. The goal of energy policy therefore has to be to bring about and maintain self-sustaining further development with other renewables. In a long-term solar scenario from 1997,[8] a group of experts forecast that some €39 billion would be needed by 2010, but the energy costs that would be offset amount to €29 billion. In other words, some €10 billion in start-up financing would save around €700 million per year over 14 years in this scenario.

At first glance, €700 million a year might seem like a lot, especially in light of current budget deficits. Yet, far greater subsidies are already made available in energy policy (see Figure 11.2). In 2005, the subsidies for domestic anthracite in Germany amounted to around €2.7 billion,[9] and the tax exemption for kerosene is equivalent to some €8 billion each year.[10] In comparison, Germany's important 100,000 Roofs Programme to promote photovoltaics only cost around €100 million a year.[11]

In 2000, Germany's Renewable Energy Act (EEG) found a way to provide sufficient funding to grow renewables without being a burden on the state budget: feed-in rates. For instance, in 2004 some €3.6 billion in feed-in rates was paid.[12] The value of the power sold to the grid that year (38.5 billion kilowatt-hours) amounted to around €2.3 billion,[13] leaving us with a cost of around €1.3 billion, a figure that has been rising since. But because these feed-in rates generally automatically drop over time (systems connected to the grid at a later date receive lower compensation), the total cost of feed-in rates is expected to start dropping in around 2015.

In light of the positive environmental effects that renewables have and in light of the tremendous subsidies in conventional energy, it is not hard to justify this start-up financing.

What we still do not know is how to provide effective start-up financing for solar heat in the best way. One way could be the method described in 11.15. If renewables are to become competitive with fossil energy, ecological tax reform must be continued and improved (see 11.4).

11

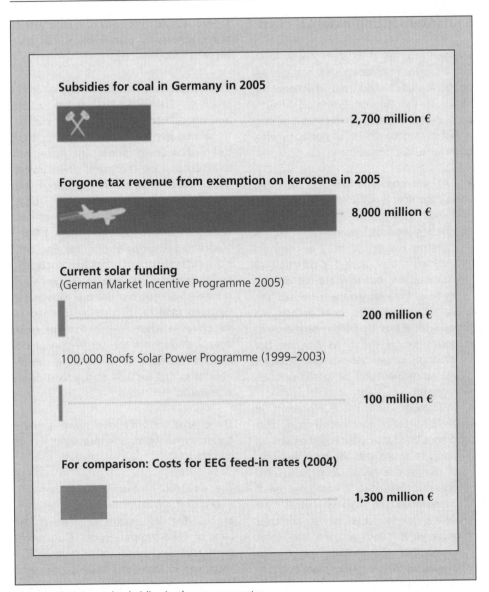

Figure 11.2 Annual subsidies in the energy sector

Source: The authors

11.3 Internalizing external costs

Energy is relatively affordable for individuals, but energy consumption can cost society dearly. A study[14] conducted by Prognos on behalf of the German Economic Ministry back in 1992 found that current pricing would eventually reduce, if not completely undermine, our prosperity.

- The environmental damage from our current energy consumption (acid rain, human health, oil disasters, nuclear disasters and the greenhouse effect) is currently not priced in. The costs are incurred either during production or consumption, but they are not covered by those who caused them. Instead, this damage – also called 'external costs' – is paid for either by certain segments of society or by society as a whole (for instance, when harvests are reduced, insurance premiums are raised or taxes increase).

- In October 2006, British economist Sir Nicholas Stern published a study that drew a lot of attention. Based on simulations, he found that the costs and risks of climate change would reduce global domestic product by at least 5 per cent if we continue with business as usual. If we take a wider spectrum of risks and their consequences into account, the report found that the damage might even increase to 20 per cent or more of global domestic product.

Clearly, energy is sold at prices below what it should actually cost. As a result, more energy is consumed than is necessary; more investments in efficient appliances and renewable energy would be made if external costs were included in energy prices.

Indeed, the external costs of human activity not contained in market prices can even exceed production costs.

Back when Prognos was working on its study, the Fraunhöfer Institute for Systems and Innovation Research found that conventional coal plants entail enormous costs not included in power prices. The researchers assumed that the coal plants under investigation have scrubbers in compliance with the standards that still apply today. Specifically, they found that a kilowatt-hour caused societal costs ranging from 1.6–6.5 eurocents per kilowatt-hour; this amount is not included in power prices. In other words, power generated with fossil energy causes some €4.5–19 billion of damage each year in Germany, and those costs are covered by society as a whole.[15] What kind of sense does it make from the perspective of our national economy to sell energy cheaply upfront so that we have to pay for it dearly afterwards?

The external costs of nuclear power are even greater when we include their societal costs and risks.[16]

The exclusion of external costs not only means that too much energy is consumed, but also that investments are made on the basis of these incorrect prices. Our current energy supply strategy, with its allegedly low costs, can in fact entail much higher total costs for society than a solar economy, which may have greater internal costs at the moment, but will always have very low external costs.[17] If we stick to our current price system, we will simply be continuing this large-scale mismanagement. Germany has taken a step in the right direction with its Ecological Taxation Reform and the Renewable Energy Act, but we have certainly not gone far enough.

Why Officer Adam dreams of the Sun

Officer Adam has an unenviable task. Once again, he has to protect a train transporting nuclear waste. But this time, the effort requires more police officers than ever, and the event will cost the public around 50 million euros.

Yet, the police would not need to be tied up with such things if we used solar energy instead. Solar energy is environmentally friendly and cannot be exhausted. Most importantly, it does not produce radioactive waste.

For the price of one shipment of nuclear waste, we supply 100,000 square meters of solar panels. Enough to provide Officer Adam and 4,000 of his colleagues with electricity – for several generations.

The time is right for the Sun. The reasons are also obvious. Get involved.

Figure 11.3 External costs of nuclear energy

Source: Solar-Fabrik AG

171

11.4 Ecological taxation reform – protecting jobs and the environment

In most cases, electricity and heat from renewable sources are still more expensive than conventional power and heat for various reasons:

- The external costs of the consumption of fossil fuels are not included in the price (see 11.3).
- Some of the technologies for renewables are still being developed, and large-scale production – which would bring costs down – is not yet ready.
- When renewables are used, new technologies replace fossil energy. This renewable technology largely entails manual labour, which is very expensive in Germany and other industrial countries, partly because it is highly taxed.
- Fossil energy and nuclear energy continue to receive generous subsidies (see 11.1 and 11.2).

This is where ecological taxation reform comes in. A tax on fossil and nuclear energy and other raw materials with environmental impact makes the use of these materials more expensive, thereby indirectly making the production of efficient equipment and renewable energy less expensive in comparison.

A switch from fossil energy to renewable energy and efficient technologies results in a number of positive effects:

- Toxic emissions and emissions of heat-trapping gases (especially CO_2) drop when less fossil energy is consumed.
- Our dependence on fossil energy, which is very great in Germany and in the EU as a whole, is reduced.
- Because the growth of renewables creates more jobs than are lost when less coal, oil and natural gas are consumed (see 10.5), jobs are created.
- Lower unemployment rates reduce the social cost of unemployment when the state does not have to spend as much on unemployment benefits, does not lose so much tax and Social Security revenue, and does not have to cover illness brought about by unemployment.

One common misconception about ecological tax reform (ETR) is that it raises the overall tax level. In fact, the very concept of ETR is that the tax level be kept the same. The goal is to shift taxation from labour, which we want to have and is available in sufficient quantities, to scarce resources, which we want to use efficiently. Furthermore, properly designed ETR also ensures that the poor are not detrimentally affected, which was unfortunately not the case in the first stages of ETR in Germany.

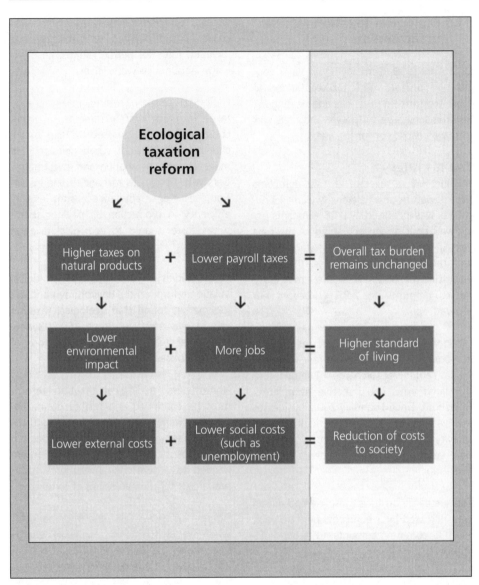

Figure 11.4 How ecological tax reform works

Source: The authors

11.5 Ecological taxation reform in increments

On 1 April 1999, Germany began its ecological taxation reform. A tax was added to fuels, electricity, natural gas and heating oil, and the surcharge for power and fuels was increased each year for five years.

The tax rates

Starting with a tax rate of 1.02 eurocents per kilowatt-hour on electricity (April 1999), the tax was increased by 0.26 eurocents per kilowatt-hour at the beginning of the next year. In January 2003, the last stage began with the fourth tax hike. Overall, the tax on electricity for households and small businesses amounts to 2.05 eurocents per kilowatt-hour.

Up to the end of 2002, businesses involved in production, agriculture and forestry paid a 20 per cent lower surcharge.[18] Energy-intensive large firms could receive even larger reductions. Since 1 January 2003, businesses involved in production, agriculture and forestry have paid 60 per cent of the total rate, equivalent to 1.23 eurocents per kilowatt-hour.

The surcharge on fuels (gasoline and diesel) was increased by 3 eurocents per litre each year; it currently amounts to around 15 eurocents per litre.

There is no tax on natural gas and heating oil which is burned in cogeneration units[19] to promote power generation in such units.

Overall, some €20 billion in tax revenue was collected and devoted to stabilizing and lowering rates for pension funds, i.e. non-wage labour costs were offset.

Ecological taxation reform raised energy taxes in predefined increments, allowing consumers and businesses to take future changes into account when decisions are made about consumption and investments. Cars with better gas mileage, home insulation, efficient appliances and energy efficiency in production all pay for themselves faster thanks to ecological taxation reform.

Three studies on the effects of ecological taxation reform on the environment and the job market found that ecological taxation reform was effective. Some 250,000 new jobs were created, and 20 million tons of CO_2 was offset.[20]

Nonetheless, ecological taxation reform must be continued and further optimized. Climate change and scarce fossil resources are worsening faster than expected. Additional stages of ecological taxation are needed to ensure that investments today prepare us for the necessities of tomorrow.

Figure 11.5 shows the trend in demand for oil if we switch to a renewable energy system over the long term. Without ecological taxation reform, demand skyrockets, but if we increase taxes on fossil energy demand drops considerably. Here, the financing for renewable energy is easier to get.

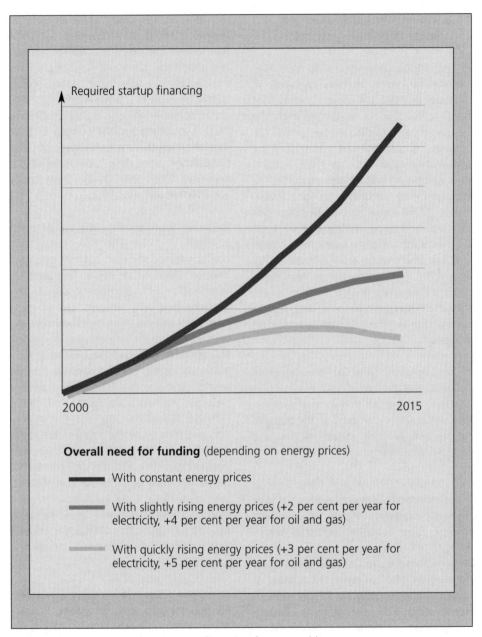

Figure 11.5 Eco-taxes reduce start-up financing for renewable energy

Source: Langniß et al, 1997

11.6 Investment bonuses for solar thermal systems

In the 1990s, widespread support was offered for solar thermal systems in Germany. By 1997, the country had installed nearly 200,000m^2 of collectors, with the German states providing 77 per cent of the bonuses.[21] Surveys showed that most of the system owners would not have installed their systems yet without this support.[22] The bonuses clearly led directly to these investments. At the same time, these campaigns also clearly ramped up the market; most German solar installers said they installed their first solar thermal arrays between 1992 and 1996.[23] Overall, these campaigns reached their goals. A large number of new solar thermal systems were installed, technical expertise became widespread, and prices dropped.[24]

In the light of types of governmental budgets, the design of efficient support strategies for the future is especially important, as we see below based on the example set by two German states: Hessen and Baden-Württemberg.

From 1992 to 1995, both of these German states provided investment bonuses for solar thermal systems. Baden-Württemberg offered about DM2000 deutsche marks (€1022) per single-family home, whereas Hessen offered up to DM3000 (€1534, but not exceeding 30 per cent). The number of systems installed grew by 77 per cent in Hessen and 193 per cent in Baden-Württemberg.[25] Clearly, this policy led to great market growth in those years.

In 1996, things changed when special federal funding was provided for solar thermal systems on new buildings. Hessen responded by changing its support; to compensate for the lower system costs, support was cut by a third, and funding was concentrated on existing buildings. Despite these two restrictions, the number of solar thermal arrays that took advantage of these bonuses increased by an additional 14 per cent from 1995–1997. Overall, this strategy was efficient and successful.

In Baden-Württemberg, the government changed in 1996. The new centre-right coalition turned the solar bonuses into low-interest loans that could be used in combination with the federal bonus. The outcome was very different. Up to then, most solar investors had been homeowners who wanted to do something good for the environment with their savings and took advantage of the bonus. Loans were of no interest to them. The loans were therefore only used for new buildings in combination with the federal funding, and the volume was low. As a result, the solar market plummeted. From 1995–1997, the collector area installed with this funding dropped by 54 per cent (roughly down to the level of 1992, see Figure 11.6), while the number of solar thermal arrays nationwide grew by more than 40 per cent.[26] This politically motivated switch pulled the rug out from under the solar sector.

At present, Germany provides investment bonuses for solar thermal systems from its Market Incentive Programme.

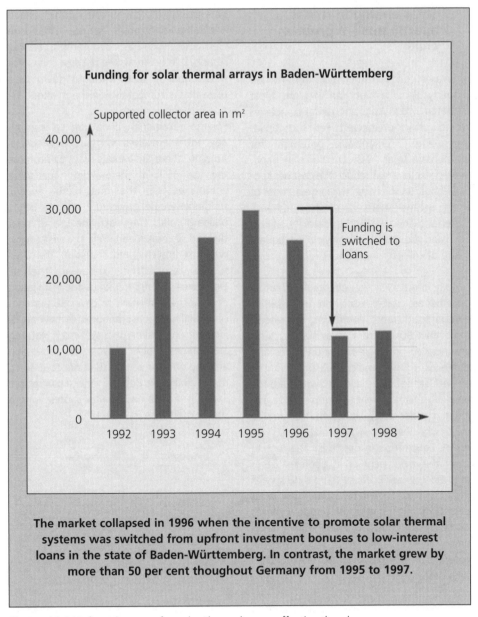

Funding for solar thermal arrays in Baden-Württemberg

The market collapsed in 1996 when the incentive to promote solar thermal systems was switched from upfront investment bonuses to low-interest loans in the state of Baden-Württemberg. In contrast, the market grew by more than 50 per cent thoughout Germany from 1995 to 1997.

Figure 11.6 Upfront bonuses for solar thermal more effective than loans

Source: Landtag Baden-Württemberg, 12/1840, 3635

11.7 Solar energy in rental apartments – a problem child

A solar strategy restricted to user-owned property does not go far enough. Most multi-family dwellings and rental complexes would remain unaffected. Yet, such buildings offer tremendous potential for inexpensive solar heat. On the one hand, large collector areas reduce the cost per kilowatt-hour; on the other, the large number of users means there is always demand. Systems then have better capacity utilization, and the useful energy yield per square metre of collector increases.

Despite these clear benefits, solar thermal systems are rarely used on multi-family dwellings and rental complexes; at present, more than 90 per cent of all such systems are installed on single-family dwellings and duplexes. One main reason is the owner/user conflict: building owners are the ones who have to make the upfront investment in solar arrays, but tenants are the ones who benefit when utility costs go down. Landlords can only raise the rent to cover the investment, as is done for other renovation work. Because there is no special legal basis for such renovations, few landlords take advantage of the option. Instead, landlords and housing management firms simply do without solar energy, leaving the potential of solar power largely untapped on such properties.[27]

Solar heat contractors are one option. Here, landlords/housing managers have the service company install and operate the solar array (and possibly the heating system). Based on the difference between investment costs and operating costs, the contractor then calculates the price for heat that can be passed on to the tenant. This business model gets around the owner/user conflict and is becoming a common service offered in Germany, such as by Berlin's Energy Agency.[28] In addition, the price of solar heat must continue to be brought down to a price that is competitive on the market.

For the transitional period until solar arrays pay for themselves without any special funding, there are various ways to promote the use of solar thermal and apartment complexes. On the one hand, special funding can be provided;[29] on the other, building codes can make the use of solar thermal systems mandatory. The first option was first implemented in Berlin. The city's senate resolved in its Solar Thermal Ordinance of 1995 to require the installation of solar thermal arrays to cover 60 per cent of annual heat consumption in all new buildings with central heating systems.[30] Housing associations fought against the regulation and the government decided not to enforce the ordinance.[31] But the city of Barcelona, Spain, has since had greater success with a similar ordinance (see 11.14).[32]

11

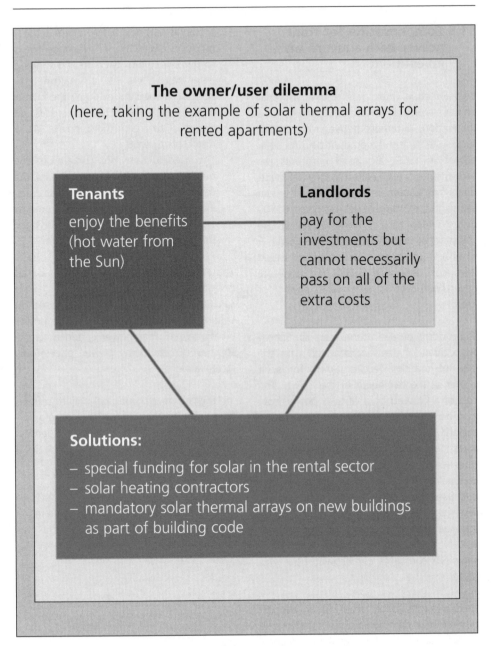

Figure 11.7 The problem child: Solar energy for tenants

Source: The authors

11.8 Compensation for solar power with a return on investment

Before feed-in rates were offered for solar power starting on 1 April 2000, there was a tremendous difference between the cost of such systems and the amount paid for solar power. In 1999, around 8 eurocents per kilowatt-hour was paid for one kilowatt-hour. If array owners sold all of their power to the grid, they would get just under €1500 over 20 years for a 1 kilowatt-peak system (20 years at 900kWh/year at 0.08 €/kWh = €1440). But back then, such a system would have cost around €10,000 and therefore never have even come close to paying for itself.

To provide a proper return, the Solar Energy Association of Aachen came up with the idea of offering feed-in tariffs for solar power at the beginning of the 1990s. The cost of a kilowatt-hour of solar power from a properly installed array would then provide a slight profit margin in addition to covering investment costs and operating costs. The rates were specified at the time of grid connection for a period of 20 years. And because the cost of new solar power systems was dropping (see 3.5), newly installed systems would receive slightly lower feed-in tariffs. At the end of the 1990s, some 90 eurocents per kilowatt-hour was paid.

Feed-in tariffs differ from other support programmes common at the time in one crucial respect: the funding does not apply to the installation of the system itself, but rather to the power produced. This system has a number of clear advantages:

- Money is only paid if the system actually produces electricity. Here, the incentive is to get the system back up and running whenever there is a malfunction. Because owners have to pay the investment costs themselves, there is an incentive to keep them down. As a result, prices drop.
- The funding does not come from the public budget; rather, utilities pass the costs on to all power consumers, insuring that the programme remains implemented when politicians look for things to cut from the budget.

The principle behind feed-in tariffs is based on one already applied by utilities, who calculate the retail electricity rate based on various costs of different kinds of power plants. When solar power is added to the sources of electricity, power prices only increase slightly.

In 1999, some 20 municipal utilities offered feed-in tariffs. None of the large conglomerates did. Figure 11.8 shows that feed-in tariffs for solar power gave photovoltaics a tremendous boost in the beginning.

The basic idea behind feed-in tariffs is to give investors in solar power systems a return on their investment, an idea that was later adopted in the Renewable Energy Act when it was revised in 2004.

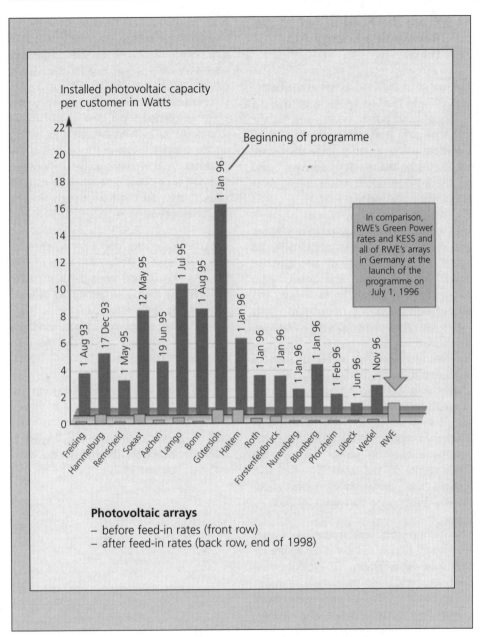

Figure 11.8 Feed-in rates for solar power get photovoltaics started in Germany

Source: Solarförderverein Aachen

11.9 From the Feed-in Act to the Renewable Energy Act (EEG)

The rollout of the Feed-in Act at the beginning of 1991 marked a turning point in the history of wind power in Germany. For the first time, grid operators were required by law to pay a floor price for renewable power sold to the grid. The rates paid were based on the average power prices for all retail customers in the previous year.

Grid operators were obligated to pay 90 per cent of that figure for electricity from wind power and solar arrays, 80 per cent for power from biomass and small hydro stations with a capacity up to 500kW, and 65 per cent for power from hydro plants with a capacity ranging from 500kW–5MW.

For instance, Germany's Bureau of Statistics calculated that the average price per kilowatt-hour in 1997 was 9.4 eurocents, so the rate paid for wind and solar power in 1999 was 8.4 eurocents per kilowatt-hour. In 2000, 0.2 eurocents less was paid for a kilowatt-hour of wind power because the liberalization of the power market had brought power prices down in most areas, thereby reducing the base price for 2000.

Potential investors were unsure about what feed-in rate they would get in the future for their renewable power. To provide more clarity and hence more investment incentives for renewables, feed-in rates were made independent of the retail rate when the law was revised.

Another change was brought about by the revision of the Energy Act in the version that took effect on 29 April 1999. Legislators have adopted a special rule for the Feed-in Act, stipulating that utilities no longer had to compensate for power if the renewable energy exceeded 5 per cent of the utility's total power sales, with the obligation falling on the supplier between the utility and the producer. And if the supplier in-between also had 5 per cent renewables in its total sales, it also did not have to pay for any additional power.

This rule – called the 'duel 5 per cent ceiling' back then – was introduced in order to limit the financial burden on individual power providers. By then, a large number of wind turbines had been installed on the northern German coast, so power distributors in these areas were more affected by the law than other power companies.[33]

But this ceiling was not a good way to solve the problem, and it would have slowed down the further growth of the wind market. So when the law was revised in 2000, the Renewable Energy Act did away with that rule (see 11.10).

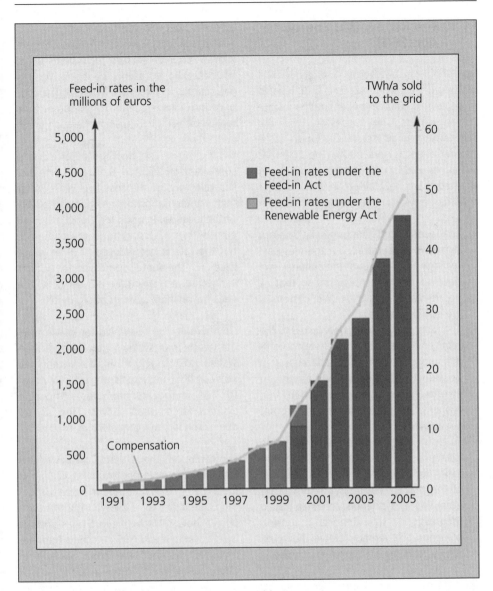

Figure 11.9 From the Feed-in Act to the Renewable Energy Act

Source: BMU, Erneuerbare Energien in Zahlen, 2006

11.10 The Renewable Energy Act (EEG)

From both an ecological and economical viewpoint, the Feed-in Act (see 11.9) needed to be revised after the liberalization of the power market. Its successor was the Renewable Energy Act (EEG), which came into force on 1 April 2000. The EEG was revised in 2004, with feed-in rates being adjusted.[34] This law differs in a number of important ways from the Feed-in Act of 1991:

- The feed-in tariffs are originally paid out by network operators, but the fees paid are later equaled out at the national level between network operators so that all network operators and power customers are equally affected.[35]
- The rates paid are no longer linked to the retail power rate, but rather vary according to the type of renewable energy in question in order to allow for a return on investments in renewable systems run properly. The rates paid remain constant for 20 years.[36]
- To ensure that the rates keep up with the falling cost of renewables, the compensation for solar arrays, wind turbines and biomass facilities decreases automatically according to a calendar schedule (called 'degression' in financial jargon); compensation remains stable, however, for 20 years once a system is installed.

The EEG has the following floor prices for facilities that went into operation in 2007:

Hydropower: Here, 9.38 eurocents per kilowatt-hour is paid for systems up to 500kW, whereas systems with an output of 501–5000kW receive a floor price of 6.45 eurocents per kilowatt-hour.[37]

Biomass: Here, compensation also depends upon the size of the system, with 10.99 eurocents per kilowatt-hour being paid for systems up to 150kW, 9.46 eurocents per kilowatt-hour being paid for systems up to 500kW, 8.51 eurocents up to 5MW, and 8.03 eurocents for larger systems.[38] The floor price increases if power mainly comes from renewable resources in a cogeneration unit.[39]

Wind power: The floor price for systems connected in 2007 is 8.19 eurocents per kilowatt-hour in the first five years. After that, compensation depends on the quality of the location. In good locations, only 5.18 eurocents is paid starting in the sixth year, but the rate is not reduced at all in worse locations. The annual degression for systems connected at a later date is 2 per cent. The rules for offshore systems also differ.[40]

Geothermal: The rates paid for power from the geothermal systems also depend upon system size, ranging from 15 eurocents per kilowatt-hour in systems up to 5MW down to 7.16 eurocents per kilowatt-hour in systems larger than 20MW. The annual degression is 1 per cent starting in 2010.

Photovoltaics: The greatest improvements have been made in photovoltaics. Instead of the 8.5 eurocents that used to be paid under the Feed-in Act for a kilowatt-hour of solar power, now rates between 51.8 eurocents (solar façade) and 37.95 eurocents (ground-mounted systems) are paid. Starting in 2006, compensation drops automatically by 6.5 per cent per year for new ground-mounted arrays.

The EEG has given investors a fair, reliable framework and led to a boom in renewable energy. At the beginning of 2010, the new governing coalition in Germany proposed drastic one-off cuts in the feed-in tariffs for solar power. For the current tariffs, see http://de.wikipedia.org/wiki/Erneuerbare-Energien-Gesetz.

	up to 150 kW	up to 500 kW	up to 5 MW	>5 MW
Hydropower[1]	9.38	9.38	6.45	special rule
Waste and mine gas[2]	7.33	7.33	6.36	special rule
Biomass[2]	10.99	9.46	8.51	8.03
	up to 5 MW	up to 10 MW	up to 20 MW	>20 MW
Geothermal[3]	15.00	14.0	8.95	7.16
	up to 30 kW	up to 100 kW	>100 kW	
Photovoltaics on roofs and noise barriers[4]	49.21	46.81	46.30	
as solar façades[4]	54.21	51.81	51.30	
ground-mounted[5]	37.95	37.95	37.95	special rule

No project ceiling

Wind power[6]		
Onshore	8.19	for at least five years (max. 20 years),
Offshore	9.10	5.18 c/kWh afterwards for at least 12 years, special rule

Automatic annual rate reductions for systems connected after 2007 in %	[1] *1.0*	[4] *5.0*
	[2] *1.5*	[5] *6.5*
	[3] *1.0*	[6] *2.0*
	(starting in 2010)	

Figure 11.10 The Renewable Energy Act: Feed-in rates for new systems connected to the grid in 2007 (eurocent/kWh)

Source: The authors

11.11 The EEG as a model for other countries

Germany's Renewable Energy Act (EEG) has not only brought about a boom in renewable power in Germany, but also led a number of countries to copy Germany's success in their own legislation; not only German wind power and solar power equipment, but also the policy itself has become a hot export item.

In addition to practically all of Europe – from Spain to Portugal, France, Austria, Ireland, The Netherlands, Greece, Italy and the UK – countries such as Thailand and Pakistan have also adopted this model, as have regions such as Ontario, Canada and the US state of Vermont. China also now offers feed-in rates for renewable energy, including biomass and solar energy.

A growing number of countries are adopting feed-in rates. At the beginning of 2006, the Global Status Report found that 41 countries and states had adopted feed-in rates.[41] Often, these feed-in rates are part of a package of policy instruments.[42] In 2009, at least 64 countries now have some type of policy to promote renewable power generation.[43]

Since 2010 Britain also has a feed-in tariff. The programme, like the successful programmes it was modeled after, was designed to 'set tariffs at a level to encourage investment in small-scale, low-carbon generation.'[44]

The global boom in renewable energy can be expressed in system capacity or, perhaps even better, in dollars and cents. Total investments in renewables only amounted to US$30 billion in 2004, but that figure had increased to US$120 billion in 2009. Renewables have clearly become a sector worth billions, and the industry is growing far faster than other industries.

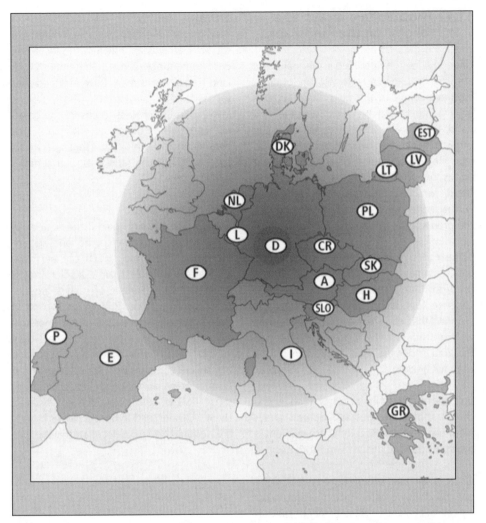

Figure 11.11 German feed-in rates abroad: Countries with feed-in rates for renewable electricity

Source: The authors

11.12 Photovoltaic arrays as a 'blight' on the landscape

Under the feed-in rates paid in Germany, it is attractive for investors to set up ground-mounted arrays in the field. Such systems offer a number of benefits.[45] On the one hand, there are economies of scale; on the other, solar cells can be oriented optimally to face the sun, and the modules do not heat up as much when they have air behind them, which increases overall efficiency. Furthermore, maintenance is easier to do on the ground than on roofs or façades.

Some concerns about nature conservation are unfounded. For instance, the stands used to install the panels do not have concrete foundations and do not seal the soil. And the shade cast by the panels moves with the sun just like it does with trees, so the land under the solar panels can still be used for grazing.

But one major bone of contention is the charge that large ground-mounted arrays are a blight on landscapes. The argument is similar to the one against wind turbines.[46]

German law was therefore revised in 2004 to provide a greater incentive for solar power on buildings, where plenty of space is still available, but field systems are still possible. From 2004–2009, roughly 12 eurocents per kilowatt-hour more was paid for building-integrated solar arrays than for ground-mounted systems (see 11.10). In spring 2010, the German government proposed to completely do away with feed-in tariffs for ground-mounted arrays on farmland; in addition, support for arrays on brownfields would also be drastically cut.

Siting

Feed-in rates are only available, however, if the solar array is approved in the local land development plan. Since 1 September 2003, land development plans have included sites set aside for solar arrays. The public is involved in the process to ensure the greatest possible public acceptance. One consequence of this policy is that solar arrays cannot be built on land against a community's wishes.

Site restrictions

Feed-in rates are only paid if the solar array is on disused or contaminated land, including disused farmland, which can be made ecologically valuable again if a solar array is installed.[47] One example of such a ground-mounted array is the 3.4MW solar plant in Borna, Saxony, where 22ha of a former coal briquette plant was converted (see Figure 11.12).[48] Since German law does not provide any special compensation for solar arrays on normal green areas, much less ecologically sensitive land, solar arrays are not installed there. Without feed-in rates, systems in such areas would never pay for themselves.

In 2006, a 3.44MWp solar array was installed on the grounds of a former coal briquette plant in Borna

Figure 11.12 Solar on disused land

Source: Geosol Gesellschaft für Solarenergie mbH

11.13 Quotas and requests for proposals

There are generally two policy designs to promote renewable power production. The first is floor prices for power sold to the grid (see-in rates). Here, the market decides what the volume is; if the floor price is high enough, a lot is invested, but if the floor price is too low, little or nothing is invested.

The second type of policy does not set the price, but rather the volume in quotas or requests for proposals.

When quotas are imposed, utility companies are obligated to get a certain percentage of their power sales from renewable sources. Power generators get certificates to demonstrate that they have sold a certain unit of renewable power (such as a certificate for 10,000kWh). Companies that produce more than the quota requires can sell certificates to those who do not produce enough. The price for the certificates depends on supply and demand. Quotas require a regulatory body, which monitors the issuing of certificates and imposes penalties if quotas are not fulfilled.

In requests for proposals, the policy stipulates a target volume, such as additional annual wind power production of 300GWh. Investors can then submit proposals for some of that volume, and the least expensive bidders get the contracts.

Proponents of such policies consider them more efficient than feed-in rates because they are allegedly more competitive, and only the cheapest projects go online.

Are they right? Without a doubt, they are in theory. But in practice, volume-based policies have more often discouraged than encouraged the growth of renewables.

While Denmark and Spain have had a grand success expanding their renewables markets with floor prices, the UK, Ireland and France did not go anywhere with their quota policies, which is why they have all switched to feed-in rates.

Although we cannot compare these two policy types in detail here, a decade of data clearly shows that countries with floor prices (feed-in rates) installed several times more wind power in 2004, for instance, than countries with volume targets.

The German state of Baden-Württemberg – roughly the size of Connecticut and with similar solar conditions – alone had 1074MW of solar online at the end of 2008, producing around 1 billion kilowatt-hours of electricity. In comparison, the US had a total installed PV capacity of only 800MW at the end of 2008, a full 25 per cent less than tiny Baden-Württemberg – a fact that the US solar sector likes to hide by claiming: 'Installed solar power capacity in the US rose by 17 per cent to 8775MW in 2008.'[49] However, that figure includes all kinds of solar, including pool heating, etc., none of which is included in the figure for Baden-Württemberg.

11

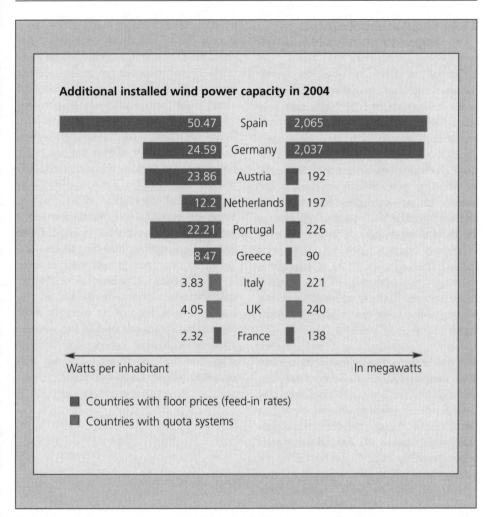

Additional installed wind power capacity in 2004

50.47	Spain	2,065
24.59	Germany	2,037
23.86	Austria	192
12.2	Netherlands	197
22.21	Portugal	226
8.47	Greece	90
3.83	Italy	221
4.05	UK	240
2.32	France	138

Watts per inhabitant   In megawatts

■ Countries with floor prices (feed-in rates)
■ Countries with quota systems

Figure 11.13 Quota systems and feed-in rates: Which is more effective?

Source: Bundesverband WindEnergie

191

11.14 Solar thermal arrays required on new buildings

In Germany, a Market Incentive Programme has promoted solar thermal arrays and wood pellet heating systems over the past few years. Rising oil prices brought about tremendous growth here, but the growth also put a strain on public budgets (which were financing these investment bonuses). Nonetheless, renewable heat still did not grow as fast as renewable electricity, which was supported by feed-in rates. For instance, the share of renewables on the heat market only grew from 5.1–5.4 per cent from 2003–2005, while the share of renewable electricity rose from 8.1–10.2 per cent.[50] Faster growth is therefore needed in the heating sector if we are to enter the Solar Age.

One way to step up this transition is to make solar thermal systems mandatory on new buildings. In Germany, the small town of Vellmar just outside of Kassel was the first to take this step. In 2001, the Osterberg neighbourhood required solar thermal arrays in its urban planning code.[51] Some 1000m^2 of collector area had to be installed on approximately 350 new buildings. Hamburg later went even further in its Climate Protection Act, which required solar collectors in urban planning. Some 5500 apartments now get 30 per cent of their hot water from solar thermal systems.

In Spain, Barcelona went much further. Since August 2000, the Solar Ordinance[52] has required all residential complexes with at least 16 apartments to get at least 60 per cent of their hot water from solar thermal arrays. Swimming pools even have to get 100 per cent. This rule does not apply for a single neighbourhood, as in the German examples, but rather all over the city. In the beginning, the construction sector opposed the ordinance, but there is now widespread acceptance of the mandatory solar arrays. Since the end of 2000, installed collector area has grown more than tenfold.[53] Other Spanish communities, including Madrid and Seville, quickly followed suit, and in October 2006 a nationwide ordinance took effect with somewhat lower requirements: all new buildings and renovation projects must ensure that 30–70 per cent of hot water is provided with solar energy. The national obligation does not override stricter ordinances in individual communities.

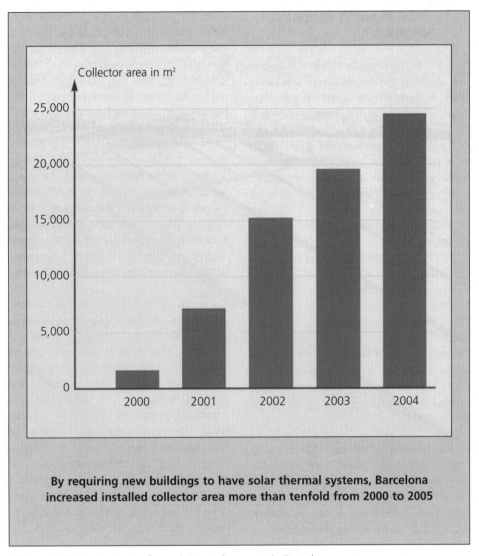

Collector area in m²

By requiring new buildings to have solar thermal systems, Barcelona increased installed collector area more than tenfold from 2000 to 2005

Figure 11.14 The success of mandatory solar arrays in Barcelona

Source: Quilisch and Peters, *Solarthermie in Katalonien*, 2005

11

11.15 Feed-in tariffs for heat in Germany?

In 2005, the new coalition under the leadership of Chancellor Angela Merkel agreed to focus on the market potential of renewable heat by continuing the Market Incentive Programme with the addition of other policy instruments such as a Renewable Heat Act.[54] In May 2006, Germany's Environmental Ministry produced a consultation paper listing four possible models:[55]

1 Investment bonuses.
2 Tax incentives.
3 Models of usage (with or without a rule for compensation).
4 The Bonus Model (feed-in tariffs for heat).

The first two are basically well known policy instruments that depend upon the current budget situation. It is therefore unclear whether they will be able to provide the necessary continuity in policy support given strong growth.

In the 'models of usage', operators of heating systems and heating networks are obligated to get a certain share (such as 10 per cent) of the heat they use or market from renewables. If that is not possible or economically feasible, they must apply for exemption. Israel and Spain have already proven the success of this approach (see 11.14).

The drawback of this approach is the paperwork. In addition, large systems and district heating networks, which tend to be less expensive, also receive less support.

In the Bonus Model (feed-in tariffs for heat), generators of renewable heat use the heat themselves or sell it to third parties. For the heat they generate, they receive a bonus specified by law on top of the price consumers usually pay for fossil energy; the bonus covers the additional cost of renewable heat. The value of the bonus depends upon the technology used.

Sellers of fossil fuels have to cover these bonus payments according to their share of the market; they pass on these extra costs as a surcharge on the oil and gas they sell (see Figure 11.15). Operators of small systems receive an upfront, one-off investment bonus instead of ongoing bonus payments.

The bonus model does not depend upon the state of the public budget and can therefore provide reliability. By tailoring the amount of the bonus to what each technology needs, special incentives can be created for different types of district heating networks. The public sector plays a minor role; for instance, it determines market shares and prevents fraud. The bonus payments could be financed with a slight surcharge on the price of fossil fuels. The German Environmental Ministry estimated that this surcharge could amount to around 1 per cent (0.65 eurocents per litre of oil) in 2010 and that figure will only rise to 2.5 per cent by 2020.[56]

Overall, the bonus model provides benefits for long-term expansion, and somewhat less paperwork is involved than in the model of use. As of 2010, the concept had not been implemented in Germany.

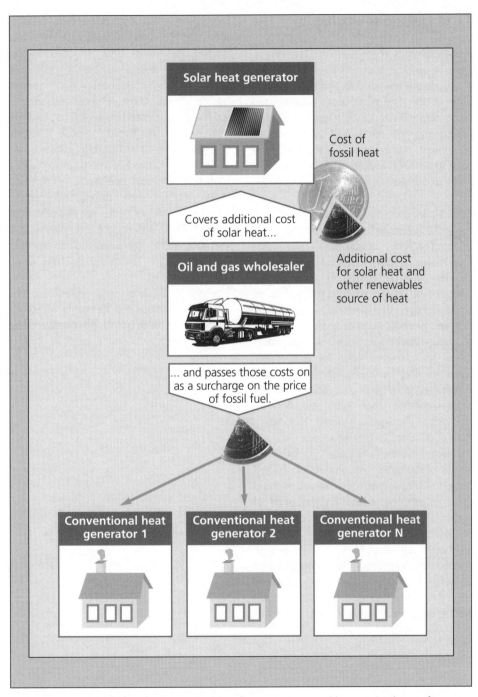

Figure 11.15 Proposal for feed-in rates for Germany: Heat providers receive bonus for environmental service

Source: The authors

11.16 Emissions trading

A distinction is made between two basic types of policy instruments in environmental policy: on the one hand, we have command-and-control approaches, which tell companies how much they can emit, what exactly they can emit, and what kind of equipment may be used; on the other, we have free-market instruments, such as taxes, levies and emissions trading.

The idea behind emissions trading is to find the least expensive way of reducing emissions of carbon dioxide (CO_2) and other heat-trapping gases. Here, a government decides what the maximum carbon emissions can be among emitters (currently only large energy companies) to protect the climate. These companies then either have to reduce carbon emissions in their own plants so they can reach the goal themselves, or they have to purchase 'emissions certificates' from other companies that have more certificates than they need. Because the overall emissions volume is clear, a market price for carbon emissions results. If the target is very ambitious, the price for carbon certificates will be higher than if companies can reach the target without great investment. Based on the price of carbon, emitters will decide whether it is cheaper to change something in-house or purchase certificates from third parties. In theory, supply and demand allows carbon emissions to be reduced in the cheapest way here. Emissions trading is therefore considered to be economically efficient.

Emissions trading began in Germany in 2005 after the EU had adopted its Emissions Trading Directive in 2003, which obligated EU member states to begin emissions trading.[57] The EU has insisted on emissions trading because there was a general consensus that the EU would otherwise not fulfill its obligations in the Kyoto Protocol, in which the signatories agreed to reduce their emissions from 2008 to 2012 to a certain target that differed from one state to another. Overall, the EU agreed to reduce its carbon emissions by 8 per cent over the reference year of 1990. Within the EU, the various member states have agreed to various reduction targets; Germany's is 21 per cent.

The Kyoto Protocol was ratified by the necessary majority and went into effect on 16 February 2005; it is binding under international law.[58]

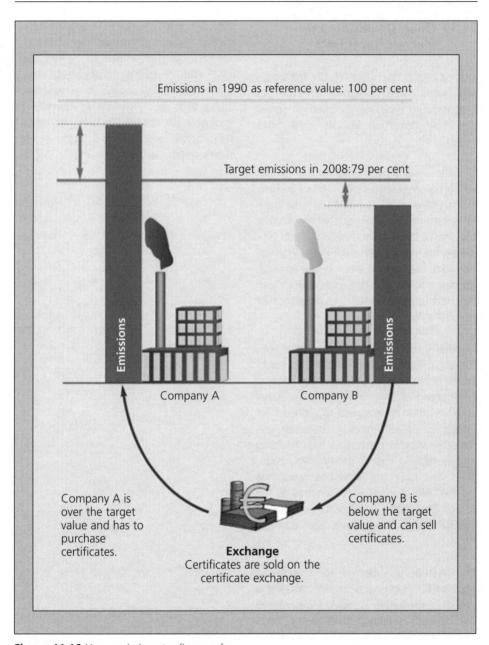

Emissions in 1990 as reference value: 100 per cent

Target emissions in 2008:79 per cent

Emissions

Company A   Company B

Emissions

Company A is over the target value and has to purchase certificates.

Exchange
Certificates are sold on the certificate exchange.

Company B is below the target value and can sell certificates.

Figure 11.16 How emissions trading works

Source: The authors

11.17 Clean Development Mechanism (CDM)

In the Kyoto Protocol (see 11.16), the signatory states agree to reduce emissions of heat-trapping gases by specified amounts. The first reduction period runs from 2008–2012.

Globally, it does not matter to the climate which country reduces its emissions by how much. Economically, it makes the most sense to reduce carbon emissions where it costs the least. To allocate funding most efficiently, the signatories to the Kyoto Protocol agreed to use 'flexible mechanisms' to reach the targets. For instance, one country can purchase emissions certificates from another to fulfill its obligations.

Another instrument that can be used both for EU obligations and within the Kyoto Protocol is the Clean Development Mechanism (CDM). Here, the industrial countries that have agreed to reduce their emissions (called Annex I countries) can instead invest in projects in industrializing and developing nations (non-Annex I countries) and have the emissions reductions credited to their own home country to the extent that these investments protect the climate and focus on sustainable development.

This system is expected to provide an impetus for investments in renewables and energy conservation in newly industrialized and developing nations. Investment projects and hydropower, wind power, reforestation (carbon sinks) and methane gas from waste thereby have another source of revenue: proceeds from the sale of carbon certificates (Certified Emission Reduction Units or CER).

Although this instrument can make a difference in climate protection, the CDM does not go far enough to help newly industrializing and developing nations achieve a sustainable energy supply. Other aspects of their energy supply systems also have to be improved, such as feed-in tariffs for renewable energy, fewer subsidies in the energy supply system, less red tape, etc.

11

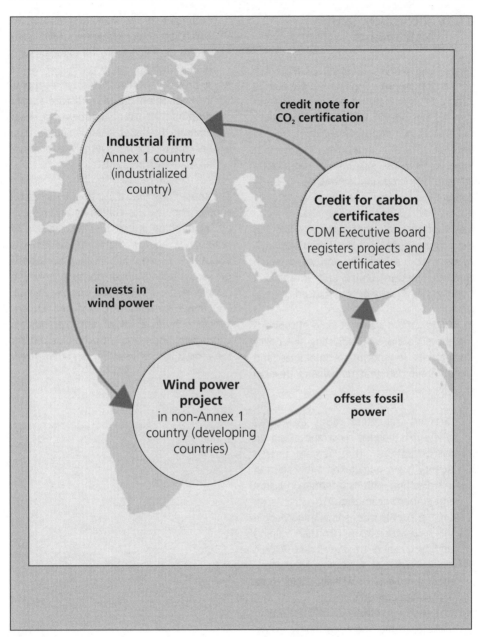

Figure 11.17 How the Clean Development Mechanism works: An example

Source: The authors

11.18 A cornucopia of instruments

Over the long term, renewable energy can and must make up a large part of our energy supply. Growth therefore has to be fast, and there is no looking back. All of the technologies at our disposal must be promoted until they are competitive. The proper approach is to use instruments that provide start-up financing until renewables become competitive on their own.

The important role that Ecological Taxation Reform has played alongside Germany's Feed-in Act, the distribution of research funding, feed-in tariffs and investment bonuses has already been discussed.

In addition, there are other ways of supporting these policies and adjusting the policy instruments to specific technologies and target groups so that renewables develop quickly.

- Targeted marketing raises awareness where it is needed most and clears up misconceptions. The German Energy Agency has already taken a few steps in this direction with its information portal and efficiency campaign.[59]
- On the supply side, some tradespeople still have reservations. Further training seminars, such as the Swiss RAVEL programme[60], and nationwide demonstration projects could help dispel some unfounded concerns.
- To ensure the efficiency of the funding made available, individual policy instruments must be adjusted both to the preferred target group and the specific technologies to be supported. Systems with a small investment volume and private users are easy to reach with investment assistance. Systems with

outputs that are easy to measure (in particular, power generators connected to the grid) can easily be paid for with feed-in tariffs.
- To provide incentives for fast investments in renewables, the rates offered should drop into the future. Furthermore, a set amount is better than a percentage because the former puts greater pressure on manufacturer costs.

The most important thing about successful support, however, is continuity and reliability. Nothing is worse for the growth of renewable energy than announcements about new policies that are then postponed or sorely underfunded, so that only the people at the front of the line are served. This effect is even worse when investment support entails a lot of red tape and is contingent upon a construction permit for the project before financing can be finalized.

	Feed-in rates	Upfront bonuses	Low-interest loans	Operating costs bonus (FITs for heat)	Risk fund
Small hydropower	✓		✓		
Wind power	✓				
Photovoltaics	✓	✓	✓		
Small solar collector arrays		✓			
Large solar collector arrays			✓	✓	
Electricity from biomass	✓		✓		
Heat from biomass		✓	✓	✓	
Geothermal	✓		✓	✓	✓

Figure 11.18 Using different policy instruments

Source: The authors

11

11.19 Phasing out nuclear

The 1986 reactor catastrophe in Chernobyl finally made everyone realize what the risks of nuclear power are. But climate change has since made it clear that our current energy supply system also poses tremendous risks. The energy sector is taking advantage of this new situation to play one risk off against the other; in the process, nuclear energy is now being sold as a way to combat the greenhouse effect. After all, the argument goes, nuclear energy offsets fossil energy, thereby reducing carbon emissions.

While correct on the surface, this claim leaves out a number of important issues. For instance, under the Schroeder government Germany resolved to decommission its nuclear plants after 32 years of service. While we are waiting for the remaining plants to be shut down, power providers have a great incentive to increase sales in order to keep future technologies from getting started. As a result, they are not telling people how to conserve energy, which would lower demand, nor are they implementing renewables themselves.

If Germany shut down its nuclear plants soon, the result would be great innovation and production in new energy technologies. The markets for sustainable energy technologies would grow quickly, which belies the claim that we would only increase carbon emissions by doing away with nuclear power. In 1996, the Institute of Applied Ecology demonstrated in a phase-out scenario that there would only be a temporary increase in emissions if nuclear were phased out completely, but that increase would quickly be compensated if power production facilities are completely revamped. Indeed, even if all of Germany's nuclear plants were switched off within a year, four years later overall carbon emissions would be back at their original level[61] provided that Germany implements an energy policy focusing on greater efficiency and renewables.[62]

Phasing out nuclear power plants would therefore speed up the transition to the Solar Age. And there would be another benefit – the risk of nuclear energy would be done away with without increasing the greenhouse effect.

Nuclear energy is a dying branch of power generation. Over the past 15 years, more plants have been taken off-line than have been connected. And that trend will presumably continue in years to come. On liberalized energy markets, power providers can hardly be expected to be interested in building nuclear plants for economic reasons. Such plants are simply too expensive up front and take too long to complete. On closer inspection, it turns out that large energy conglomerates actually do not want to expand nuclear energy, but merely keep the ones already in operation running – despite the considerable risks.

While a quick phaseout of nuclear power would increase carbon emissions in the short term, emissions would actually drop even faster than under the status quo over the mid to long term if an energy transition policy focusing on efficiency and renewables were pursued. A recent study conducted by EUtech and Greenpeace confirms these findings.

The results are taken from a study conducted by the Institute of Applied Ecology based on figures from 1996

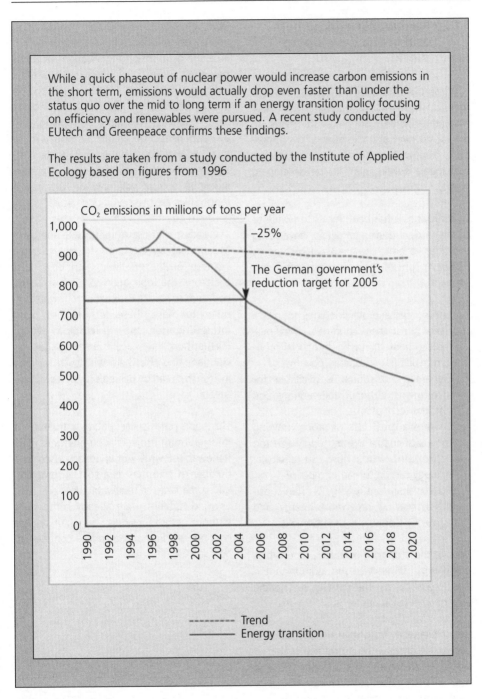

Figure 11.19 Phasing out nuclear

Source: Öko-Institute, 1996, EUtech and Greenpeace: Klimaschutz: Plan B, 2007

11.20 Renewable energy here and in the developing world

In previous sections, we have focused on the main instruments and steps towards a sustainable energy supply in Germany. In places, we have also mentioned how important renewable energy can be for sustainable development in the developing world.

As difficult as it has been for us to switch to solar energy, it is even harder for developing countries to switch to a sustainable energy supply in light of their far worse economic situation.[63]

- Environmental protection does not play a major role in these countries, which face more urgent economic and social problems, such as widespread poverty.
- Generally, no capital is available for investments in renewable energy and efficient technologies.
- There is a great lack of proper training for a sustainable energy supply and too little expertise about how a sustainability strategy can be turned into policy.
- For social reasons, energy is often subsidized, making renewable energy and efficiency technology unaffordable.

Despite these upfront obstacles, renewables and energy efficiency are just as important in these countries on the path to sustainable economic development.

Often, electricity from Solar Home Systems is cheaper in areas without grid access than it would be to expand the grid (see 3.3 and 3.4). In such cases, solar energy can quite easily improve standards of living for the poorest of the poor at the same time as it marks a major step towards sustainable development. However, for psychological and economic reasons the success of renewable energy in off-grid applications depends on the success of such applications connected to the grid. Otherwise, renewable energy will be stigmatized as a 'technology for the poor', preventing it from being used everywhere and becoming a crucial part of energy policy. An economically feasible strategy that creates jobs and opens up new industrial sectors will only be possible if renewables are used on a large scale.

The widespread use of renewable energy in the developing world is still not possible because of the relatively high cost. It is therefore the obligation of industrialized countries to reduce the cost of renewables in order to make these technologies more interesting for the developing world. Furthermore, the experience gained in Germany and elsewhere with certain policy instruments can be passed on to these countries.

This approach is successful, as a number of international projects demonstrate. The Renewable Energy Act is already used in a number of countries as a sort of template, and some types of renewable energy – such as wind turbines – are already competitive without any subsidies as their prices continue to drop and oil prices continue to rise (see 6.6).

Figure 11.20 Renewables for off-grid applications

Source: triolog, Frauenhöfer ISE, Solar-Fabrik AG

12 Good Marketing – Successful Projects

12.1 Everyone loves the sun

As we have seen in previous chapters, renewables have only been able to conquer a limited number of market segments, such as solar heating for outdoor swimming pools (2.1) and multi-family dwellings, wind power in good locations, and Solar Home Systems in the Third World (3.3). In other cases, proper policies (especially feed-in tariffs) and changes in the general business environment are needed to attain the required growth levels. Wherever solar energy is not yet competitive with other energy sources, it pays to conduct marketing campaigns to raise awareness about the additional environmental benefits of expanding renewables.

In this chapter, we present a number of examples. All of these successful marketing strategies share one thing: a focus on the widespread acceptance of solar energy and other types of renewable energy in a large majority of the population. Back in the 1970s, German politicians responded to the first oil crisis by arguing that we should move away from our dependence on a single energy source to have a broader mix consisting of nuclear energy, coal, oil, gas, and renewables. Today, there is almost no support for nuclear energy, and fossil energy hardly plays a role in the forecasts for future energy supply. A majority of the population wants to have their future energy come from renewables (see Figure 12.1).[1] A great majority (74 per cent) even wants the current support in Germany to be maintained.[2] A slightly smaller share of the population, but nonetheless a clear majority (around 70 per cent), is willing to pay more for electricity from renewables.[3]

Along with the public's willingness to pay more, the positive image that renewables have is clearly utilized in the examples discussed in the rest of this chapter. For instance, people are clearly willing to:

- Pay more for green power, just as they are willing to pay more for organic food (see 'Green electricity' 12.2, Solar brokers 12.4).
- Invest in solar, wind and biomass plants (see 'Community systems', 12.3).

Combinations of completely different products – such as shares in solar arrays and season tickets to football matches – have also already proven successful on the market (see 12.9).

The examples discussed are not exhaustive. Rather, they are intended to serve as starting points, possibly even leading to new marketing ideas. In fact, we would be interested as the authors to hear about your new ideas.

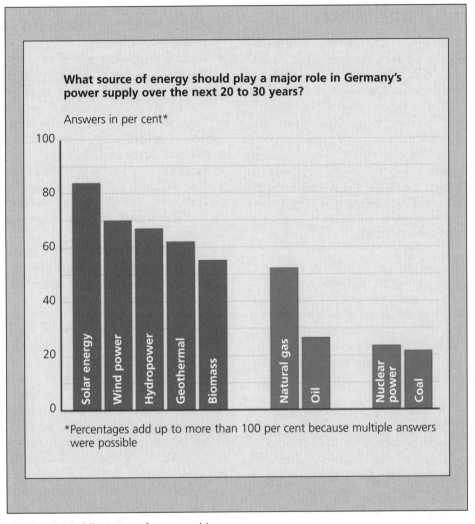

Figure 12.1 Public support for renewables

Source: Forsa, 2005

12.2 'Green electricity'

On 29 April 1998, Germany's new Energy Industry Act went into force. Since then, all power customers have been able to choose their power provider and type of electricity. Companies responded quickly by starting to offer 'green electricity'. Some even believe that the liberalization of the power market is a first step towards the energy transition described by the Institute of Applied Ecology; they believe that consumers are voting with their feet by moving away from nuclear energy and towards renewables at their own wall sockets.

What are we to make of the concept of 'green electricity'? How does it differ from good old electricity? First, we must keep in mind that the electricity itself is exactly the same in both cases. The only difference is how the power is produced. So what are the most important things that make electricity 'green'?[4]

- To begin with, 'green electricity' should come from renewables to the greatest extent possible. Other types of technologies, however, such as distributed cogeneration, also considerably reduce the environmental impact and can be a legitimate part of a green electricity package.
- More importantly, the production facilities that make this 'green electricity' should be newly installed. If old equipment is used, then consumers are merely being asked to pay more for systems that were already up and running anyway. If subscribers of green power do not force companies to install additional systems, the overall environmental impact of power generation does not improve. The only thing that does improve is the power provider's bottom line; after all,

the same electricity was already sold at a lower price to the general customer base.

- Because the product that green power customers purchase is identical to normal electricity, customers must be certain that the qualifications for this label are properly enforced. In other words, we need reliable certification and labels for 'green electricity'.
- Furthermore, we have to make certain that the amount of green power sold does not exceed the amount generated; such review should be a part of certification.
- Other criteria also offered in the 'green electricity' packages, such as 'full service' and simultaneous generation,[5] are, however, not as important.

'Green electricity' is a matter of trust. Generally, providers want to show that they are a part of a 'sustainable energy sector'. But customers nonetheless have to take a closer look at what they are supporting. Some companies that offer green electricity actually originally filed suit against the Feed-in Act. Others continue to support nuclear power strongly. In addition, some municipal power providers still do their best to prevent cogeneration units from being connected to the local grid. Such companies are merely offering 'green electricity' to improve their image; otherwise, these companies are clearly pursuing business practices that damage the climate.

In other words, even a product label can only cover certain aspects of a company as a whole.[6]

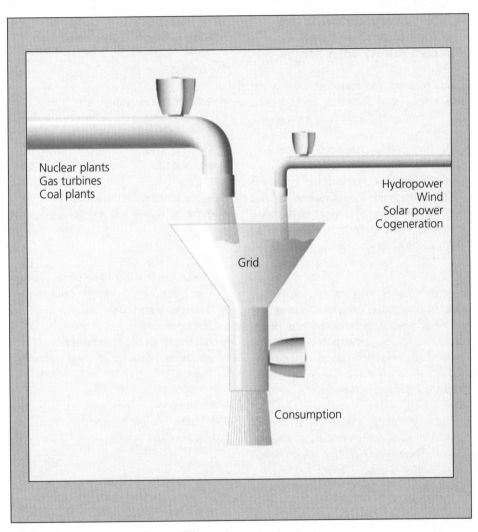

Nuclear plants
Gas turbines
Coal plants

Hydropower
Wind
Solar power
Cogeneration

Grid

Consumption

Figure 12.2 Green power from utilities

Source: The authors

12.3 Not everyone owns the roof over their head – community solar arrays

Financing is not the only obstacle towards renewables. For instance, a lot of people have the money but don't have a good roof:

- Roughly half of Germans do not own their own homes.
- Just under 20 per cent of the rest live in apartments with shared roof space, and PV arrays generally cannot be installed on such roofs without the consent of all the owners, which often proves to be an obstacle to investments in practice.

To make it easier for such people to start using solar and tap unused potential, the idea of community solar arrays was developed. Essentially, private investors buy shares of arrays set up in good locations.

There are a number of benefits:

- The community arrays are set up in optimal sites, which insures good solar yield. There is no lack of space on industry roofs and public buildings.
- Because of the system size, the specific costs are low.
- By breaking down the cost of a large array into a large number of shares, people with average incomes can also get involved.

Three groups/partners have to come together in a community solar project:[7]

The owners

- pay for the investment and
- receive compensation (minus trusteeship and overhead).

The solar firm

- plans and installs the array and
- provides service and maintenance.

The management

- draws up the installation and maintenance contracts on behalf of the owners,
- oversees installation until the final bill has been paid,
- applies for public funding and
- manages provisions and allocates income.

A number of such community projects have also come together for wind power, hydropower and biomass.

12

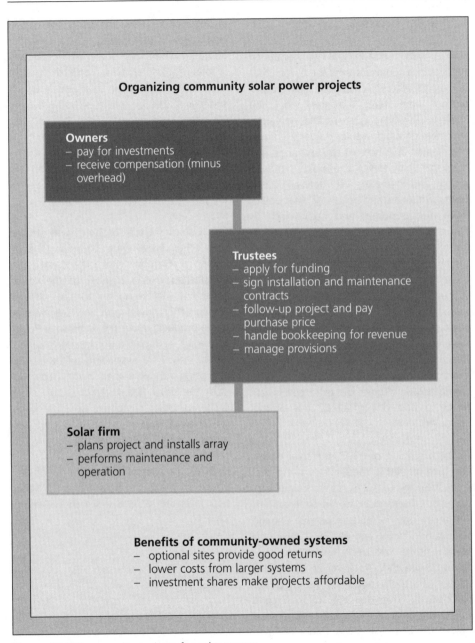

Figure 12.3 Community projects for solar power

Source: The authors

12.4 Solar brokers

Back in 1995, the Municipal Utility of Zürich conducted a representative survey of 3500 randomly selected customers. Seven per cent of them stated that they would be interested in paying more for solar power. To serve that demand, a solar power exchange was created in Zürich. At the exchange, the municipal utility serves as a broker between supply and demand, i.e. between solar power producers and buyers of solar power (both private homes and businesses). To ensure that the cheapest solar power is traded, the municipal utility issued a request for proposals for solar power. The utility then chose the 12 least expensive providers/of solar arrays from the 21 bids tendered. The average solar price from these bids was 1.2 Swiss francs per kilowatt-hour. The utility then signed a power purchase agreement with the solar power producers for 20 years. If demand for solar power were to drop during this timeframe, the extra costs were to be passed on to all of the utility's customers.

At the beginning of 1997, the solar power was then offered to these customers at cost — at the price of 1.2 Swiss francs per kilowatt-hour. Demand exceeded expectations. The utility was not able to provide enough solar power. By the end of 1997, the supply of solar power was being rationed to keep up with demand. As a result, the utility started another round of requests for proposals to cover demand. In this second round, the price of a kilowatt-hour of solar power fell below 1 Swiss franc, bringing down the overall cost to 1.11 Swiss francs per kilowatt-hour starting in 1999.

In September 1999, the utility had 5,690 customers purchasing just under 800,000kWh of solar power under contract. A total of 33 arrays with a collective output of 1.1MW supplied the solar power. By the summer of 2003, that figure had grown to 68 arrays with an output of 2.3MW. This successful model was copied in other parts of Switzerland and abroad. It quickly became a role model at more than ten power providers in Switzerland, and some of the projects were very successful.[8]

The Municipal Utility of Zürich was pleased that other firms were copying its Solar Power Exchange. But the firm also complained that a number of the copies were not close enough to the original: 'When other power providers sell watered down products under the name Solar Power Exchange – which we unfortunately cannot patent – they are misleading customers and damaging not only their own image, but also the idea itself. Unlike our stock exchange, solar customers on some other similar exchanges are simply financing solar arrays that had already been built.'[9]

By 2010, the price of a kilowatt-hour of solar power on the Zürich Solar Power Exchange had fallen to 0.70 Swiss francs (without VAT).[10]

Figure 12.4 Solar brokers: The example of Zürich

Source: Elektrizitätswerke Zürich (CH)

12.5 Service brings in new customers for all-in-one packages

The installation of a solar thermal array is quite a task for homeowners. You have to get information, compare technical concepts, choose one, see what it would cost, compare the offers, pick someone to do the job, find out about what public funding is available, apply for it and then keep an eye on your system for decades to make sure it is properly functioning – quite a lot of tasks these days, when people do not have a lot of time. Here, municipal utilities could step in and perform all of this as a service to win over some new solar friends.

For instance, utilities can offer their customers turnkey solar arrays at a fixed price. The price would include consulting services, planning and applications for public funding. To the extent necessary, the utility could even work with local contractors. The utilities could issue calls for tenders and have qualified contractors perform the work. In this way, the utility would do all of the work for the investor up to the point where the array goes into operation. Homeowners would then pay a fixed price for an array that will provide hot water practically for free for the next 20 years.[11]

In Germany, the municipal utility of Ettlingen (40,000 inhabitants) offers such an all-in-one service to both homeowners and businesses. Launched in 1996, the programme installed more than 100 solar thermal systems within the first two years, a figure that had risen to more than 400 by the end of 2005. By 2003, the utility had generated some €2.6 million of additional orders for local contractors. Customers are the first to benefit, but satisfied customers are also good for the utility because they do not move to the competition. In this respect, promoting solar energy is a good idea for the utility.

The utility also offers another service for renewable heat. Even though it is a natural gas supplier itself, the municipal utility of Ettlingen also supplies its customers with wood pellets of only the best quality to ensure smooth, low-emission operation – and certainly also to retain customers.

12

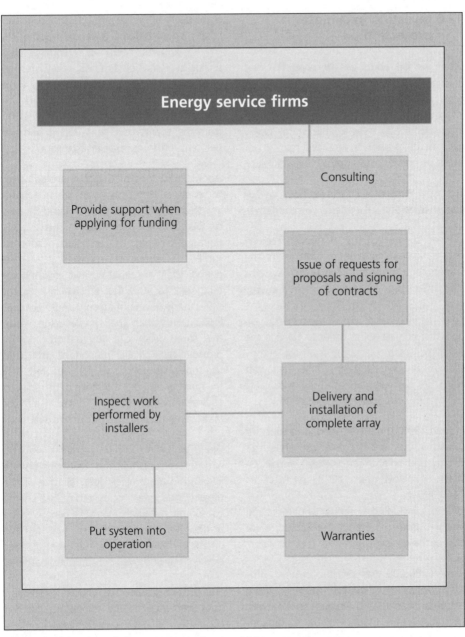

Figure 12.5 Service wins over new customers: All-in-one packages

Source: The authors

12.6 Investing in climate protection

When we talk about climate protection, we are not only talking about solar energy, but about energy in general. While it is nice to imagine how all buildings will one day benefit from solar energy, a sustainable energy supply will not be possible unless we reduce our excessive energy consumption dramatically.

The ECO-Watt project shows how energy conservation and solar energy work together.

Initiated by the Institute of Applied Ecology and developed in cooperation with the Fraunhöfer Institute for Solar Energy Systems (ISE), the project aimed to prove that energy conservation already pays for itself a lot of the time. The idea was to exploit the conservation potential in a school in order to finance the installation of a solar array from the energy savings. In July 1999, the first 'negawatt plant' in Germany financed by means of community contracting went into business.

In June 1998, the project began looking for investors via two local firms: Freiburg's Energy and Solar Agency (Fesa) and ECO-Watt. By November 1998, a total of €245,000 had been collected. Some 100 investors, including a number of parents and teachers from the Staudinger School, were part of the project.

Some €280,000 was invested in new lighting, controlled circulation pumps, more efficient heating and ventilation controls, two solar arrays, modern demand management and water conservation. The work was done by a local contractor in cooperation with a subsidiary of ECO-Watt dedicated to this project. The city of Freiburg paid the firm the difference between the old and the new energy costs for a period of eight years.

Eight years after the project started, it can clearly be considered a success. Some 1.5 million kilowatt-hours of electricity, 5.5 million kilowatt-hours of heat and 77 million litres of water were saved. Investors were not the only ones to benefit from the savings; the school also received €79,000 from the energy savings. Investors had a 6 per cent return on their investments.[12]

About 2650 tons of CO_2 was avoided over the contract term. This means someone who invested €5,000 in the project saved 53 tons of CO_2 over the eight years. This roughly equals the average CO_2 emissions accounted for by a German citizen over a five-year period. There are other positive environmental impacts as well. The efficient fluorescent tubes in the new lighting contain 90 per cent less mercury than their predecessors. Also, the same level of illumination is now provided by a smaller number of lights that use tubes with longer service lives, reducing the number of tubes that have to be replaced each year by more than 75 per cent – a major improvement for the school caretakers.[13]

Along with the technical planning, teachers and the parents/teachers association were involved in the project. After all, one of the project's goals was not only to reduce energy consumption, but also to make people aware of how they waste energy so that the new, efficient technology would be further bolstered by improved consumption patterns.

The project shows that climate protection does not have to be expensive; on the contrary, it already pays for itself in many cases. Solar thermal energy was used, as was, to a lesser extent, photovoltaics. Profit maximization was not the main goal – investing in a sustainable energy future was. And that goal can only be reached when energy conservation is combined with solar energy.

12

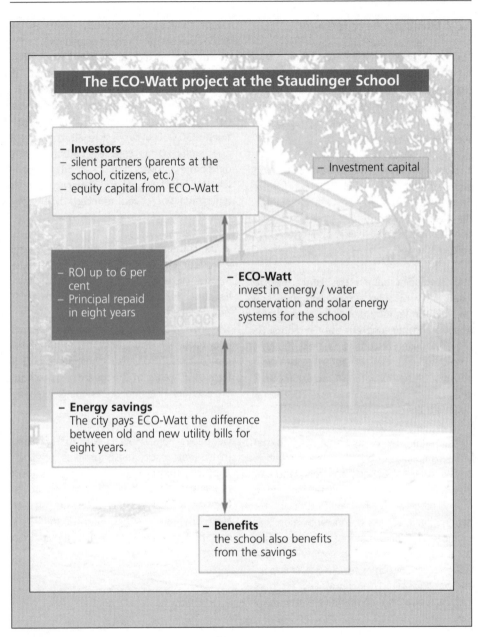

Figure 12.6 Climate protection as a good investment

Source: The authors

12.7 Utilizing new leeway

The liberalization of the power market (see 1.10) has also brought about some positive changes. For instance, compensation for power from solar arrays and small cogeneration units (up to 50kW) has been considerably improved.

In light of these changes, the ECO-Watt project described in section 12.6 was further developed. In cooperation with Ö-Quadrat, the Wuppertal Institute came up with a concept for schools called the 100,000W Solar Initiative.[14]

The basic idea behind this concept is to install 50W of solar power per pupil and save 50W in lighting demand per pupil at a select group of schools in the German state of North Rhine Westphalia. Overall, 100W of conventional power would then be offset per pupil. A school with around 1000 pupils would then represent a 100,000W solar negawatt plant.

Furthermore, the project managed to get utility companies onboard. In the renovation of the Aggertal School, the renovation of the lighting system, the installation of a 43kW solar array, hydraulic support for heating circuits[15] and pump renovation was financed with investments from community contracting. The power supply for the Aggertal school comes from a 50kW cogeneration unit, which also supplies a school with heat at a price competitive with natural gas.

Overall, the school currently produces far more electricity than it consumes. The energy conservation measures brought the original annual power consumption of 120,000kWh down to around 65,000kWh. The solar array produces some 35,000kWh per year; the cogeneration unit, an additional 230,000kWh. As a result, the school now emits 70 per cent less carbon than it did before renovation. In a second project at the Willibrord School in Emmerich, carbon emissions were even reduced by 85 per cent.[16]

In the summer of 2003, these two projects were featured in the Future Energy initiative in the state of North Rhine Westphalia because of the example they set.

The investors, most of whom are parents and teachers at the school, get more than 6 per cent return on their investments over a term of 20 years. In the meantime, 4 school projects are ongoing under the solar&save label.[17]

Germany's Renewable Energy Act was one thing that made these projects possible; another thing was the Cogeneration Modernization Act. In addition, ecological taxation reform has increased the price of electricity, but cogeneration units are exempt from the tax on gas, providing a better return on investments.

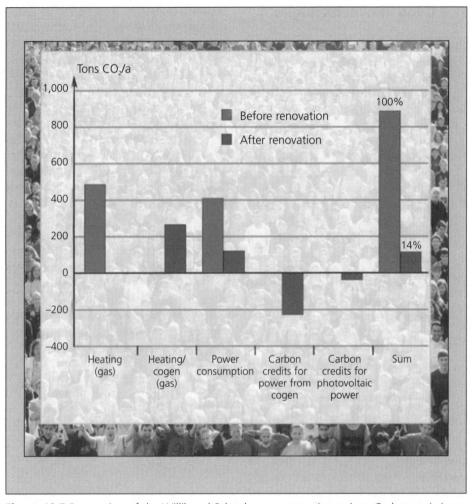

Figure 12.7 Renovation of the Willibrord School as a community project: Carbon emissions were reduced by 85 per cent in 2005

Source: The authors

12

12.8 Using new technologies

Worldwide, some 1.6 million people do without electric light today – more than in the 1880s, when Thomas Edison invented lightbulbs.[18] These people make do with oil lamps and candles to light their homes, an option that is not only expensive, but also dangerous to the environment and hazardous to people's health.

- Oil lamps give off a lot of soot, which builds up indoors and can lead to respiratory diseases and cancer. Furthermore, a low level of light makes indoor work less efficient and impairs vision over the long term.
- Because oil lamps give off very little light, the carbon emissions of this lighting technology are relatively high – roughly 20 times higher than conventional lighting with electricity from coal and oil plants. If lighting instead comes from solar power, the CO_2 emissions from oil lamps can be reduced by some 80kg per year.

Electric lighting therefore clearly reduces the environmental impact even as it improves standards of living. In regions with no grid access, solar power and compact fluorescent lighting ('Solar Home Systems', see 3.3) are currently the least expensive way of providing lighting to households. Nonetheless, the high cost keeps them from becoming more widespread.

Recently, white light emitting diodes (LEDs) were developed and will soon further reduce the cost of solar lighting considerably, which will make solar lighting available to more people. Because white LEDs make do with so little electricity, both the solar panel and the battery can be smaller, reducing the overall system cost. In addition, LEDs are harder to break and have a service life of up to around 100,000 hours, roughly 10 times longer than compact fluorescent bulbs and 100 times longer than conventional light bulbs.

The cost of efficient solar LED lighting with a focusing lens is already clearly lower than the cost of conventional lighting, such as oil lanterns, candles and battery-operated torches.[19] Efficient technology and solar energy can therefore bring light to remote regions of the world with funding coming in the form of microcredit.

12

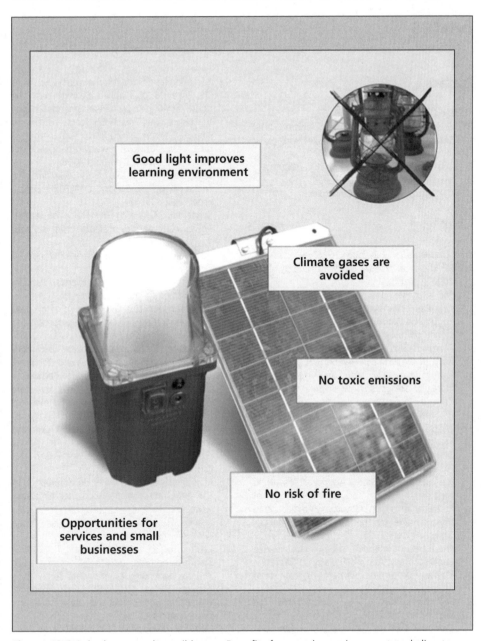

Figure 12.8 Solar lamps replace oil lamps: Benefits for people, environment and climate

Source: The authors

Notes

Preface

1   *Die Zeit*, 2 November 2006, p26.
2   German Solar Industry Association (2006) *Statistical Figures of the German Solar Industry*, June 2006.
3   *Renewables Global Status Report*, 2009, p9.
4   *Neue Energie*, March 2010, p94.
5   *Die Zeit*, 2 November 2006, p26.

Chapter 1

1   *Handelsblatt*, 13 October 2006.
2   BP Statistical Review of World Energy, June 2006.
3   For example, US oil specialist Professor Kenneth P. Deffeyes declared 24 November 2005 to be World Peak Oil Day, the day when global oil production peaked. Taken from *Energiedepesche*, March 2006, p.4. See also www.peak-oil.de and Heinberg, R. (2004) *The Party's Over: Oil, War and the Fate of Industrial Societies*, New Society Publishers, Canada.
4   *BP Statistical Review of World Energy*, June 2006.
5   *Statistisches Jahrbuch für die Bundesrepublik Deutschland 2006*, p467.
6   For example, the European Commission and the International Energy Agency (IEA) warned their member states about the unreliability of Russian energy giant Gasprom as a reaction to threats made by the firm's CEO. *Financial Times*, 20 April 2006.
7   This list is not intended to be exhaustive. See also Institute of Applied Ecology (2005) *Risiko Kernenergie: Es gibt Alternativen!*, Institute of Applied Ecology, Freiburg.
8   See 'Geisterflug auf der Geisterbahn', *Die Zeit*, 10 August 2006.
9   Germany is by no means alone in Europe in doing so. Denmark, Greece, Ireland, Italy, Luxembourg, Austria and Portugal have renounced nuclear power. In 1984, Sweden resolved to phase out nuclear, and Spain implemented a moratorium to prevent the construction of additional nuclear plants in the same year.
10 Institute of Applied Ecology (2005) *Risiko Kernenergie: Es gibt Alternativen!*
11 The Institute of Applied Ecology puts the figure 'at 40 years, whereas Greenpeace says the figure could reach 65 years. Institute of Applied Ecology (2005) *Risiko Kernenergie: Es gibt Alternativen!*; www.greenpeace.de/themen/atomkraft (February 2006).
12 Rechsteiner, R. (2003) *Grün Gewinnt: Die Letzte Ölkrise and Danach*, Orell Füssli, Zürich, especially pp39–45.
13 Schneider, M. (2005) *The Status of Nuclear Power in the World*, presentation in Brussels, 7 November 2005.
14 Bernward Janzing in *Badische Zeitung*, 10 January 2007.
15 Arbeitsgemeinschaft Energiebilanzen, January 2006, www.ag-energiebilanzen.de.
16 Bundesanstalt für Geowissenschaften and Rohstoffe (2006) *Reserven, Ressourcen and Verfügbarkeit von Energierohstoffen 2005*, Hannover. Nonrenewable energy is categorized into two groups: reserves and resources. The reserves are proven and affordable with current technology. The volume therefore depends on current energy prices. Resources have either been proven but are currently not affordable, or they have not been proven but are geologically probable.
17 Geothermal is an exception (see 7.5–7.7). It can round off solar power.
18 In addition, land and water do not heat up at the same rate, which also causes air turbulence.
19 Scheer, H. (1993) *Sonnenstrategie: Politik ohne Alternative*, Munich, p109.
20 Lehmann, H. and Reetz, T. (1995) *Zukunftsenergien: Strategien einer neuen Energiepolitik*, Birkhäuser Verlag, Berlin.
21 Krause, F., et al. (1980) *Energiewende: Wachstum and Wohlstand ohne Erdöl and Uran*, Frankfurt.
22 *Deutscher Bundestag* (1990) *Schutz der Erde: Eine Bestandsaufnahme mit Vorschlägen zu einer neuen Energiepolitik*, Bonn, especially vol 2, p158ff.
23 von Weizsäcker, E., Lovins, A. B. and Lovins, L. H. (1998) *Factor Four: Doubling Wealth, Halving Resource Use – The New Report to the Club of Rome*, Earthscan, London.
24 Nitsch, J. (2007) *Study commissioned by*

German Ministry for Environment, Nature Conservation and Nuclear Safety (BMU), February 2007.

25 Umweltbundesamt: Wie private Haushalte die Umwelt nutzen – höherer Energieverbrauch trotz Effizienzsteigerungen, Hintergrundpapier, November 2006

26 This example was used by Stryi-Hipp, G. in Sonnenenergie frei Haus - Solarthermie and Photovoltaik nutzen, presentation on 22 January 1997, Bonn.

27 Irrek, W. and Thomas, S. (2006) Der Energie SparFonds für Deutschland, Düsseldorf.

28 The amount of heat that dissipates from power plant cooling towers is greater than the heat used to keep apartments and offices warm in Germany.

29 The Verband der Elektrizitätswirtschaft (VDEW) wrote in their annual report of 1996 that they had had a survey conducted, which found that the 50 largest power providers had faced the challenge of businesses wanting to generate their own power in nearly 2000 cases from 1994–1996. With flexible pricing, the power providers were able to persuade the businesses not to generate their own power in most cases.

30 Bundesverband Kraft-Wärme-Kopplung, Stellungnahme zur Novellierung des Kraft-Wärme-Kopplung-Gesetzes, September 2006.

31 Traube, K. (2005) Potentiale der KWK. Bundesverband der Kraft-Wärme-Kopplung e.V., January 2005.

32 In least-cost planning, the cost of implementing energy conservation is compared to the cost of expanding power plant capacity and the grid. The goal is to choose the cheaper option.

33 The ecological taxation reform and the Renewable Energy Act have brought about a lot of change here.

34 Around 38 per cent in 2006.

35 Süddeutsche Zeitung, 7 September 2006.

36 Main-Spitze, 5 October 2006.

37 A private household in Freiburg with an annual consumption of 3500kWh paid €778.55 if power came from the local utility (Badenova); the cheapest offer in Germany would have been €570 (Flexstrom, 3600-Paket). The green of power offers from Lichtblick and EWS-Schönau (Watt ihr spart) were also cheaper than the local utility at €730.39 and €748, respectively. www.stromauskunft.de

38 An energy service provider not only sells power and gas, but also works to make sure that these energy sources are used as efficiently as possible, for instance by financing power conservation consulting services and campaigns for efficient power consumption.

39 See Hennicke and Seifried (1996) Das Einsparkraftwerk, Berlin, Basel, Boston.

40 Wirtschaftsminister Riehl quoted in Taz, 5 October 2006.

41 On the one hand, coal and heating oil, which have been exempt up to now, should be taxed, but the number of exemptions should also be reduced (kerosene, net-metering, refineries, agriculture, etc.). To prevent further redistribution effects, the tax rate for the production sector should be raised slightly more than the rate for private consumers.

42 Irrek, W. and Thomas, S. (2006) Der EnergieSparFonds für Deutschland, Düsseldorf

43 For example in England, Denmark and some states in the US.

44 See Wuppertal Institute (2001) Energieversorger auf dem Prüfstand, Wuppertal.

45 Energy conservation contracting is a contractual agreement between a firm (contractor) and a building owner. The contractor makes the investments and performs the work needed to conserve energy in the building. The building owner then pays the contractor the difference between the old and the new energy costs to cover the contractor's expenses. The building owner does not have to invest, does not run any risk, and nonetheless benefits from the success of renovations.

46 Part of a sustainable energy sector is that resources are not consumed faster than they can 'grow back'.

47 Just because conservation measures would pay for themselves does not mean that they are actually implemented.

48 May, H. (2006) 'Kostensenker vom Dienst', Neue Energie, October 2006, p14.

49 PROKON (2006) Der Wind hat sich gedreht, Sonderveröffentlichung, Itzehoe.

50 Joint press release of UBA and BMU on 17 October 2006 on the launch of the new Centre of Competence for Climate Impact and Adjustments.

Chapter 2

1   For further details, see Schüle, R., Ufheil, M. and Neumann, C. (1997) *Thermische Solaranlagen: Marktübersicht*, Ökobuch Verlag, Staufen bei Freiburg, p. 41ff. See also Deutsche Gesellschaft für Sonnenenergie (2006), *Nutzerinformation Solarthermie*, Munich.

2   This calculation is based on a heating oil price of 6 eurocents per kilowatt-hour (roughly 60 eurocents per litre); in addition, it is assumed that the price of heating oil will increase by 3 per cent per year. Because prices rose from 1998–2006 by a much higher average of 13 per cent, no financing costs were calculated for. Under these assumptions, the savings in the heating oil costs amount to €4800 over 20 years. If the system replaces an electric water heater, the savings amount to €10,200 over 20 years; Bundesverband Solarwirtschaft (2006) *Zeitung für Solarwärme*.

3   The share of solar arrays used for space heating rose to around 45 per cent in 2006 as a share of newly installed systems; Bundesverband Solarwirtschaft (2006) *Zeitung für Solarwärme*.

4   Thermochemical heat storage opens up new perspectives. Five times as much heat can be stored as in a conventional hot water tank of the same volume. Here, reversible chemical reactions take place; for instance, a solid can absorb a fluid. In addition to the greater energy density, one major benefit is that the heat is stored for a long time practically without any losses. For a presentation of such storage projects, see Thermochemische Speicher, BINE-Projektinfo 2/01.

5   R. Kübler and N.Fisch (1998) 'Energiekonzept für die Zukunft. Neckarsulm Amorbach – solare Nahwärme in der 2. Generation', *Solarenergie* vol 1, pp36–37; Berner, J. (2001) 'Sommersonne für den Winter: Langzeit-Wärmespeicher haben ihre Tauglichkeit bewiesen', *Sonnenenergie* Nov. 2001, pp16–19; see Rentzing, S. (2006) 'Coole Sonne', *Neue Energie*, vol 4, pp38–43.

6   Rentzing, S. (2006) 'Coole Sonne', *Neue Energie*, vol 4, pp38–43.

7   For details on the process, see Hindenburg, C., et al (1999) 'Kühlen mit Luft', *Sonnenenergie*, vol 1, pp 39–41.

8   The dew point is the temperature at which the gas part of a gas/vapour mixture (air with water vapour here) is completely saturated. The water vapour condenses when the temperature drops.

9   Hindenburg, C., et al (1999) 'Kühlen mit Luft', *Sonnenenergie*, vol 1, pp39–41.

10 Rentzing, S. (2006) 'Coole Sonne', *Neue Energie*, vol 4, pp38–43.

11 W. Mühlbauer and A. Esper (1997) 'Solare Trocknung', *Sonnenenergie* vol 6, pp35–37; Köpke, R. (2002) 'High tech by low cost', *Neue Energie*, vol 2, pp38–41.

12 In Europe north of the Alps, sludge can be affordably desiccated with solar energy in large air collectors that resemble greenhouses. Bux, M. (2002) 'Solare Klärschlammtrocknung – eine ernstzunehmende Option?', *Baden-Württembergische Gemeindezeitung*, vol 4, pp145–147.

13 Brake, M. (2006) 'Luftkollektoren heizen Wohnhäuser. Warme Luft von der Sonne', *Sonnenenergie*, September 2006, pp30–36.

14 BUND (2005) *Vorbildliche Energieprojekte von Kreisen and Kommunen in Hessen*, Frankfurt.

15 'Start für Parabolrinnenkraftwerk', *Solarthemen*, 8 June 2006, p4; Keller, S. (2006) 'Es geht voran', *Sonnenenergie*, July 2006, pp32–37; Pitz-Paal, R. (2006) *Solare Kraftwerke*, presentation at the Quo vadis Solarenergie symposium, Freiburg, 15 September 2006.

16 REN21: Renewable Global Status Report, 2009 Update, www.ren21.de. See also nreldev.nrel.gov/csp/solarpaces.

17 www.desertec.org

18 In July 2006, construction of a 1.5MW demonstration unit was begun in Jülich, Germany; *Badische Zeitung*, 5 July 2006.

19 Klaiß, H. and Staiß, F. (1992) *Solarthermische Kraftwerke für den Mittelmeerraum*, Springer-Verlag, Heidelberg; Pitz-Paal, R. (2006) *Solare Kraftwerke*, presentation at the Quo vadis Solarenergie symposium, Freiburg, 15 September 2006.

Chapter 3

1   See Leuchtner, J. and Preiser, K. (1994) *Photovoltaik-Anlagen*, Marktübersicht 1994/1995, Institute of Applied Ecology, Freiburg, p13ff, and current reports in Photon (www.photon.de).

2   Leuchtner, J. and Preiser, K. (1994) *Photovoltaik-Anlagen*, Marktübersicht 94/95, Institute of Applied Ecology, Freiburg, p56f;

Weithöner, H. (2003) 'Aus alten Zellen neue machen: Deutsche Solar steigt in Freiberg in das PVRecycling ein', *Sonnenenergie*, March 2003, pp41–43.

3 Kilowatts-peak describes the peak output of a photovoltaic array; i.e. the electricity generated under full sunlight.

4 Stryi-Hipp, G. (Bundesverband Solarwirtschaft e.V.) (2006) 'EEG-Novelle und regneratives Wärmegesetz – Chancen und Herausforderungen für die Deutsche Solarindustry', presentation, July 2006.

5 In sparsely populated rural areas, a grid connection can cost thousands of US dollars. Gabler, H. and Preiser, K. (1999) 'Photovoltaik – ein Baustein zur nachhaltigen Entwicklung netzferner Regionen', Forschungsverbund Sonnenenergie (c/o DLR), *Nachhaltigkeit and Energie, Themen 1998/1999*, Cologne, pp28–31.

6 REN21: Renewables 2005: Global Status Report, Washington 2005, p32f.

7 Schmela,M. (1999) 'Kleinvieh macht auch Mist: Wo sich in unseren Breiten Solarstrom-Anlagen rechnen', *PHOTON*, vol 2, pp44–51.

8 International Solar Energy Society (2002) *Sustainable Energy Policy Concepts (SEPCo)*, study conducted on behalf of the German Ministry of the Environment, Nature Conservation and Reactor Security, Freiburg. For further information, see www.ises.org/shortcut.nsf/to/ICNFINAL.

9 Microgrids provide power to a village not connected to the country's power grid. Within the small grid, power can be generated from fossil energy (such as a diesel engine), from solar energy, or by a combination of different technologies.

10 Professor J. Luther, director of the Fraunhöfer Institute for Solar Energy Systems until 2006, in an interview in *Bild der Wissenschaft*, vol 9, 2006, pp96–99; BINE (2005) 'Photovoltaik – Innovationen bei Solarzellen and Modulen', *Themen-Info*, vol 3, Bonn.

11 One reason was a lack of solar silicon. In the meantime, this bottleneck has been resolved, partly by the opening of production plants for solar silicon.

12 www.solarwirtschaft.de/medienvertreter/pressemeldungen/meldung.html?tx_ttnews[tt_news]=12944&tx_ttnews[backPid]=737&cHash=64a1a4abfd.

Chapter 4

1 See Steimer, G. and Lohbihler, I. (1998) *Energiekonzepte für zukunftsfähige Neubauten*, Freiburg, pp9 and 11.

2 The stipulations in Germany's Energy Conservation Ordinance (EnEV) cannot be directly compared to those of its predecessor because the new limit values applied to primary energy consumption in a building. In other words, they also take account the type of fuel used and the efficiency of the heating system. The previous ordinance only specified limit values for heating energy demand.

3 Goetzberger, A. and Wittwer, V. (1986) *Sonnenenergie*, Stuttgart, p132ff.

4 The U-value of a building component indicates how much power (how many watts) is lost over an area of $1m^2$ for every degree of temperature (1K).

5 For examples, see www.solarbau.de, www.energie-projekte. de and www.baunetz.de/infoline/solar.

6 Ebök (Tübingen) (1998) *Energieeinsparung bei Neubausiedlungen durch privat- and öffentlich-rechtliche Verträge*, Cologne.

7 Stadt Köln; Ministerium für Arbeit, Soziales... NRW: Planen mit der Sonne, Arbeitshilfen für den Städtebau, Düsseldorf, Cologne 1998, pp12 and 48: The authors estimate 0.1 euro-cents per kilowatt-hour offset.

8 The example is taken from Ranft, F. and Haas-Arndt, D. (2004) *Energieeffiziente Altbauten*, Cologne, p122ff.

9 As of March 2007; for current information, see www.kfw.de.

10 Germany's Energy Conservation Ordinance is not based on heating demand, but rather sets a limit on primary energy consumption, which also takes account of how heat is generated. The installation of a solar thermal array or a wood pellet stove can therefore help you reach the targets in Germany's Energy Conservation Ordinance.

11 One factor in determining indoor comfort is the surface temperature of walls. In poorly insulated buildings, people might feel cold even though room temperature is high enough simply because the walls are cold.

12 Russ, C. et al (1996) *Energetische Sanierung einer Typenschule in Plattenbauweise unter Verwendung Transparenter Wärmedämmung (TWD)*, Fraunhöfer Institute for Solar Energy Systems, annual report, p18.

13 Switchable glazing is another option. If hydrogen or oxygen is in between the panes of glass, a thin layer of wolfram oxide will turn

either dark blue or transparent. *Solarthemen*, 15 February 2001, p7.

14 Optimized multi-layer polycarbonate boards can also be used for this purpose. BINE (1996) *Transparente Wärmedämmung zur Gebäudeheizung*, profi info 1/96, Bonn.

15 Fachverband Transparente Wärmedämmung e.V., www.fvtwd.de.

16 Germany's Energy Conservation Ordinance stipulates the limit on primary energy consumption in buildings, not heating energy consumption. Depending on the technology used, 15kWh/m^2*year (the passive house standard) is equivalent to 20–40kWh/ m^2*year). A KfW-40 building (the required energy standard for funding from Germany's KfW Bank) is roughly equivalent to the passive house standard. BINE (2003), themeninfo II/03, *Energieeffiziente Einfamilienhäuser mit Komfort*, Bonn, p2.

17 Cost Efficient Passive Houses as European Standards (CEPHEUS). Under project number BU/0127/97, this project was funded as part of the EU's 5th Framework Programme and jointly conducted by Germany, France, Austria, Sweden and Switzerland.

18 Köpke, R. (1999) 'The next generation: Passiv-Solar-Häuser setzen neue Maßstäbe beim Energiesparen', *Neue Energie*, vol 4, pp34–35.

19 Witt, J. and Leuchtner, J. (1999) *Passivhäuser: Nischenprodukt oder Zukunftsmarkt? Eine Marktpotentialstudie,* presentation at the 3rd Passive House Conference 19–20 February 1999, Bregenz.

20 IG Passivhaus press release on 6 August 2005; www.ig-passivhaus.de

21 BINE (1994), *Energieautarkes Solarhaus*, project number 18, Bonn 1994; FhG ISE (1992) *Das energieautarke Solarhaus*, Freiburg.

22 Today, this building is used by the Fraunhöfer Institute for Solar Energy Systems as a research and institute building.

23 www.solarsiedlung.de.

24 This was the original concept, but the solar thermal array was left out for reasons of cost.

Chapter 5

1 Fritsche of the Institute of Applied Ecology puts the potential at 700PJ; Kaltschmitt puts it at 940PJ; Fritsche, U. (2004) *Bioenergie, Nachwuchs für Deutschland*, Institute of Applied Ecology; 'Biomasse – Potenziale in Deutschland', *Energiedepesche*, vol 2, 2004, p18.

2 Some 4.4 million hectares (roughly 25 per cent of our current agricultural land) of energy crops could provide 1200PJ, equivalent to around 9 per cent of primary energy demand. Fritsche, U. (2004) *Bioenergie, Nachwuchs für Deutschland*, Institute of Applied Ecology. The Fachagentur Nachwachsende Rohstoffe (FNR) puts Germany's biomass potential at 17.4 per cent of the country's primary energy consumption, a figure somewhat higher than Fritsche's. FNR (2005) *Basisdaten Bioenergie Deutschland*, Gülzow.

3 Thomas, F. and Vögel, R. (1989) *Gute Argumente: Ökologische Landwirtschaft*, Beck'sche Verlagsbuchhandlung, Munich.

4 Fachverband Biogas e.V.

5 Fachverband Biogas press release on 1 February 2007.

6 FNR (2005), *Basisdaten Biogas Deutschland*, Gülzow.

7 Bensmann, M. (2007) 'Freie Fahrt für Fermenter', *Neue Energie*, vol 1, pp52–55. Bensmann, M. (2005) 'Mächtig Gas geben', *Neue Energie*, vol 2, pp50–52.

8 Dederer estimates that a 180kW biogas unit would need 77ha, at 13 tons of dry corn per hectare. For grass (8t TM/ha), a full 137ha would be needed. Dederer, M. (2005) *Wirtschaftlichkeit von Biogasanlagen*, presentation at the Biogas-Tagung der Akademie für Natur- and Umweltschutz, Stuttgart, 19 October 2005.

9 BMU (ed) (1999) *Erneuerbare Energien and Nachhaltige Entwicklung*, Bonn, p56.

10 Böttger, G. (2010) 'Wachstumsmarkt Holzenergie', *Sonnenenergie*, January/ February 2010, p26ff; Sowie, *Neue Energie*, vol 2, p74. At www.ECOTopTen.de, the Institute of Applied Ecology presents especially environmentally friendly wood pellet heaters.

11 The Sunmachine creates heat with wood pellets, and a Stirling engine then converts part of the heat into electricity, www.sunmachine.de.

12 Forstabsatzfonds (2003) *Holzenergie für Kommunen*, 3rd edition, Bonn, p10.

13 Manufacturers report that 50–70 per cent of pellet heaters are sold in combination with solar thermal systems. Riedel, A. (2007) 'Ein starkes Tandem', *Sonnenenergie*, March 2007, pp20–22.

14 KEA (1999) *Faltblatt Klimaschutz durch Holzenergie*, Karlsruhe.

15 Forstabsatzfonds (2003) *Holzenergie für Kommunen*, 3rd edition, Bonn, p88ff.

16 Neumann, H. (2006) 'Feinstaub kratzt zu Unrecht am Holzimage', *Moderne Energie & Wohnen*, vol 1, pp67–70; Bensmann, M. (2006) 'Schreckgespenst Feinstaub', *Neue Energie*, vol 7, pp48–51.

17 Lewandowski, I. and Kaltschmitt, M. (1998) 'Voraussetzungen and Aspekte einer nachhaltigen Biomasseproduktion. Akademie für Natur- and Umweltschutz Baden-Württemberg: Biomasse', *Umweltschonender Energie- and Wertstofflieferant der Zukunft* , vol 27, pp19–38.

18 Alt, F. (1999) *Schilfgras statt Atom: Neue Energie für eine friedliche Welt,* Piper.

19 'Hier keimt die Power', *Neue Energie*, vol 4, 2006, pp47–49.

20 Schreiber, E. (2005) 'Schweden gibt Biogas!', *Solarzeitalter*, vol 3, pp25–26.

21 *Chancen and Perspektiven flüssiger Bioenergieträger in Deutschland*, conference report of 14 February 2003, Böblingen. Newsletter March 2003 from the Biomasse Info-Zentrums (BIZ), Stuttgart.

22 In 2005 some 1.3 million hectares of rapeseed was planted in Germany, roughly 727,000ha of which was used to produce biodiesel. Approximately half of that was on disused land. Information based on personal contact with Dr A. Weiske, Institut für Energetik and Umwelt, Leipzig, 9 November 2006.

23 FNR (2006) *Biokraftstoffe, eine vergleichende Analyse*, Gülzow, p20.

24 Since 1 August 2006, there has been a tax of 9 eurocents per litre on biodiesel; it is expected to rise to 45 eurocents per litre by 2011. Pure vegetable oil has also been taxed since 2008. The tax on biodiesel quickly reduced sales, partly also as a result of lower oil prices. Germany's Biofuels and Renewable Fuels Association says that biodiesel was no longer competitive starting at the beginning of 2007, and 25 per cent of Germany's biodiesel production capacity has already been decommissioned or stopped; press release on 20 February 2007.

25 FNR (2005) *Basisdaten Biokraftstoffe*, Gülzow; *Neue Energie*, vol 2, 2006, p35.

26 *German Bundestag* 14/8711.

27 FNR (2006) *Biokraftstoffe, eine vergleichende Analyse*, Gülzow, p22.

28 Das Bioethanol wird nicht direkt verkauft, sondern dem Benzin beigemischt (bis 10%) oder zum Oktanverbesserer ETBE umgewandelt und dann dem Benzin beigemischt. Bensmann, M. (2009) 'Alkohol-Diät', *Neue Energie*, vol 5, p72ff.

29 Nitsch, M. and Giersdorf, J. (2006) *Biokraftstoffe zwischen Euphorie and Skepsis – am Beispiel Brasilien, Solarzeitalter*, vol 2, pp36–41; Sieg, K. (2005) 'Zuckersüße Alternative', *Neue Energie*, vol 3, pp81–85.

30 Flex-fuel vehicles are necessary because more sugarcane than would be used to produce sugar, thereby reducing the supply of ethanol, if sugar prices are high and oil prices are relatively low. In such cases, flex-fuel vehicles could simply run on normal petrol.

31 Nitsch, M. and Giersdorf, J. (2005) *Biotreibstoffe in Brasilien*, FU Berlin, Volkswirtschaftliche Reihe 12/2005, Berlin June 2005, p12.

32 *Neue Energie*, vol 5, 2009, p74.

33 To attain the EU's target of 5.75 per cent biofuels as a share of the petrol market by 2010, some 2.1 million tons of ethanol are needed – roughly equivalent to 1 million hectares of wheat fields or around 400,000ha of sugarcane.

34 Moser, S. (2003) 'Biomasse – Karriere eines Schmuddelkindep', *Bild der Wissenschaft*, vol 5, 2003, pp66–69.

35 FNR (2006) *Biokraftstoffe: Eine vergleichende Analyse*, Gülzow, p30.

36 For the time being, however, BTL production requires absolutely dry wood; processes based on straw are not optimal. Bensmann M. (2006) 'Den Königsweg finden', *Neue Energie*, vol 5, p63.

37 Bringezu, S. and Steger, S. (2005) *Biofuels and Competition for Global Land Use*, conference documentation from the Heinrich Böll Foundation: Bio im Tank, Berlin, p62ff.

38 EU-15 takes up some 35 million hectares of land for food production outside of its own territory, roughly twice as much farmland as in all of Germany. That figure does not include imports of animal feed. Overall, EU citizens use roughly a sixth of the land they need for food production outside of the EU. Bringezu, S. and Steger, S. (2005) *Biofuels and Competition for Global Land Use*, pp5 and 67.

39 'If the world were to adopt such consumption and supply patterns (including the 60 Mtoe biofuel equivalent), there would no longer be any relevant amount of natural grassland, pampa or prairie. Outside of forests there

would be crop land intensively cultivated by ploughing and fertilization – a green desert around the globe. Certainly a theoretical scenario. The question is to which extent it may become reality.' Bringezu, S. and Steger, S. (2005) *Biofuels and Competition for Global Land Use*, p69.

40 Klute, M. (2007) 'Fremdwort Nachhaltigkeit', *iz3w*, no 298, January/February 2007, p34; 'Biodiesel-Import in der Kritik', *Neue Energie*, March 2006, p29.

41 For example, rotating crops provides ecological benefits and greater yields in Germany. Bensmann, M. (2004) 'Im Kessel Buntes', *Neue Energie*, vol 8, pp50–51.

42 Berger, T. U. (2005) *Mobilität durch Lebensenergie*, conference documentation of the Heinrich Böll Foundation: Bio im Tank, Berlin, p5.

Chapter 6

1 'Bundesverband Windenergie', *Süddeutsche Zeitung*, 8 December 1998. In the past few years, the price of a kilowatt-hour of wind power has hardly dropped. Improved system technology, however, allows slightly more power to be generated per installed kilowatt. Erfahrungsbericht zum EEG, *Deutscher Bundestag* 14/9807, pp10–15.

2 DLR, IFEU, Wuppertal Institute (2004) *Ökologisch optimierter Ausbau der Nutzung erneuerbarer Energien in Deutschland*, Forschungsvorhaben im Auftrag des Bundesministeriums für Umwelt, Naturschutz und Reaktorsicherheit, Stuttgart, Heidelberg, Wuppertal, p164f.

3 *Süddeutsche Zeitung*, 8 August 1997.

4 For example, Baden-Württemberg's governor Erwin Teufel attempted to prevent the construction of two wind turbines in August 2003 on Schauinsland Mountain although the turbines had basically already been completed based on a construction permit. (*Landtag Baden-Württemberg*, 13/2395)

5 We realize how little space wind turbines need when we compare that space to the area required by other renewables. For instance, a modern wind turbine (1.8MW, 1800 hours running at full capacity per year) generates some 3.2 GWh per year. To generate the same amount of electricity, photovoltaics would need around 5.4ha, an area roughly 200m by 270m, with 60 per cent of the area being covered (1000kWh per 10m² of panels). If the power is generated from biomass, roughly 200ha of energy crops would be needed, roughly 40 times the area of photovoltaics. This calculation does not include the energy input for corn, for instance, but the share of heat generated is also not included.

6 BMU (ed) (1999) *Erneuerbare Energien and Nachhaltige Entwicklung*, Bonn, p22.

7 Weinhold, N. (2006) 'Die Flaute im Kopf', *Neue Energie*, vol 2, pp39–42.

8 In Denmark, this has already happened with the resolution to phase out nuclear power and the adoption of clear energy policy goals, such as volume targets for wind power.

9 www.wind-energie.de.

10 BMU (ed) (1999) *Erneuerbare Energien and Nachhaltige Entwicklung*, Bonn, p31.

11 Abo-Wind, press release on 4 September 2002.

12 *Erneuerbare Energie* vol 5, 2005, p40.

13 Neue Energie, 2/2002, p. 21

14 Exports of wind turbines are extremely important for Denmark today. The country's only export item that brings in more revenue is pigs. *Naturstrom-Magazin*, vol 1, 2001. But since the law changed in Denmark in 2001, Danish manufacturers have only been able to sell a few turbines domestically. See *Neue Energie* vol 4, 2003, p44.

15 Press release from the BMU, 16 May 2006.

16 www.wwindea.org/home/index.php?option =com_content&task=blogcategory&id=21&It emid=43.

17 www.wind-energie.de.

18 Press release from the BWE, 27 January 2010.

19 *The New York Times*, 9 October 2006.

20 Of course, economic viability also depends on the rates paid for power from onshore and offshore wind turbines.

21 Most of the systems are larger than 2MW, www.offshore-wind.de.

22 REN21 (2009) *Renewables Global Status Report*, 2009 Update.

23 www.dena.de/fileadmin/user_upload/ Download/Dokumente/Projekte/ESD/netzs-tudie1/dena-grid_study_summary.pdf.

24 http://www.dena.de/en/topics/energy-systems/projects/projekt/grid-study-ii.

25 Lönker, O. (2006) 'Flaute fällt aus', *Neue Energie*, vol 3, pp42–45.

26 'Bundesverband Windenergie: Repowering – weriger ist mehr', *Hintergrundinformation*. 2005.

27 Press release from the BWE, 27 January 2010, www.wind-energie.de.

Chapter 7

1   For an overview of turbine technology and types of hydropower, see Landesinitiative Zukunftsenergien NRW, *Wasserkraftnutzung*, Düsseldorf, no date.
2   *BP Statistical Review 2006*, www.bp.com/statisticalreview; BMU (2006) *Erneuerbare Energien in Zahlen*, Berlin.
3   WWF Deutschland (2004) *Wasserkraft – weltweit von hoher Bedeutung*, Berlin; Köpke, R. (2004) 'Auf ewig heiß', *Neue Energie*, vol 10, pp85–91.
4   Landesinitiative Zukunftsenergien NRW, *Wasserkraftnutzung*.
5   www.energiedienst.de.
6   Verband der Deutschen Elektrizitätswirtschaft.
7   Staiß, F. (ed) (2001) *Jahrbuch Erneuerbare Energien* , Stiftung Energieforschung Baden-Württemberg, Bieberstein.
8   For instance, a small firm from Karlsruhe (Hydro-Watt) found a way to get old water mills to produce as much as 30 per cent more power. The water mills used to run at constant speed because the generator required that speed for a connection to the grid. But modern electronics allows the speed of water mills to be adjusted to the actual amount of water. A frequency converter ensures that the frequency of the power sold to the grid is the right one; *Neue Energie*, vol 5, 1999.
9   The Wikipedia list of the world's largest power plants now shows nine such plants with an output exceeding 5000MW, with four others under construction. www.wikipedia.de.
10 The great criticism of hydro plants has reduced the amount of funding that the World Bank provides for storage dam projects. Nonetheless, China continues to build storage dams, and the country now has roughly half of all major storage dams in the world. Kreutzmann, H. (2004) 'Staudammprojekte in der Entwicklungspraxis: Kontroversen and Konsensfindung', Geographische Rundschau, vol 12, pp4–9.
11 *Badische Zeitung*, 31 May 2003.
12 As in biogas units, the anaerobic decomposition of biomass releases methane, hydrogen sulfide and ammonia.
13 Kohlhepp, G, (1998) 'Große Staudammprojekte in Brasilien', *Geographische Rundschau*, vols 7–8, pp428–436.
14 Blume, G. (2006) 'Eine neue Mauer', *Taz*, 22 May 2006.
15 REN21, Global Status Report 2005, p8f.
16 Sanner, B. (2002) 'Das Geothermiezeitalter wird eingeläutet', *Erneuerbare Energien*, vol 2, pp52–55.
17 The geysers produce 7784GWh of power per year, see http://iga.igg.cnr.it/geoworld.
18 www.geothermal-energy.org/210,welcome_to_our_page_with_data_for_indonesia.html.
19 In Reykjavik, pavements and car parks are kept free of snow with geothermal heat.
20 www.nea.is/geothermal/electricity-generation.
21 BINE (2001), 'Geothermie', *basisEnergie 8*, February 2001.
22 *Zukunftsenergien aus Nordrhein-Westfalen*, Düsseldorf 2002, p52.
23 This kind of geothermal power production does not use boiling water (as is common in conventional power plants), but rather a fluid with a lower boiling point (such as pure hydrocarbons). Such an ORC process can generate electricity starting at temperatures above 100°C.
24 Levermann, E. M. (2004) '"Heißer Strom" aus 2.200 Meter Tiefe', *Sonnenenergie*, May 2004, pp32–34.
25 Geothermal electricity generation in Soultz-sous-Forêts ; BINE-Projektinfo 04/2009, www.bine.info/en.
26 'Bad Urach erschließt gigantische Energiequelle', *Schwäbische Zeitung*, 27 April 2002.
27 'Am Oberrhein boomt die Erdwärme', *Badische Zeitung*, 23 December 2004.
28 Marter, H. J. (2002) 'Strom aus der Tide', *Neue Energie*, vol 11, pp60–62; (2003) 'Rotor unter Starkstrom', *Neue Energie*, vol 8, pp58–60; and (2005) 'Lange Wellen', *Neue Energie*, vol 10, pp84–88; (2004) 'Seaflow – Strom aus Meeresströmungen' BINEProjektinfo, vol 4
29 Graw, K. U. (2006) 'Stand and Perspektiven der Nutzung der Meeresenergie', *Energiewirtschaftliche Tagesfragen*, vol 3, p59ff. Starting in 2008, a Pelamis system was to provide power for 2000 homes off the Orkney Islands. The Scottish government plans to provide nearly €20 million for this project; Strom von der Seeschlange; www.energieportal24.de on 28 February 2007; www.pelamiswave.com/content.php?id=159

Chapter 8

1   BINE (2001) 'Geothermie', basisEnergie 10, December 2001.

2   Auer, F. and Schote, H. (2008) *Nicht jede Wärmepumpe trägt zum Klimaschutz bei, Schlussbericht des zweijährigen Feldtests Elektro-Wärmepumpen am Oberrhein*, Lahr 2008, www.agenda-energie-lahr.de.

3   'Wärmepumpen   durchgefallen', *Energiedepesche*, March 2001.

4   It follows that the environmental impact of hydrogen mainly depends on how it is created (with power from coal plants, nuclear power or renewable energy).

5   In catalytic combustion, hydrogen directly reacts with air in contact with a catalyst (such as platinum) at relatively low temperatures. Catalysts are therefore very efficient and have very low emissions, especially of nitrous oxide.

6   The drawbacks are just as apparent, however. If hydrogen is to be 'green', it will have to come from renewable energy. Unfortunately, because renewable electricity is still relatively expensive and so much energy is lost in conversion, green hydrogen would be even more expensive.

7   An exhibition of the Landesgewerbeamt Baden-Württemberg and the Deutsche Forschungs- and Versuchsanstalt für Luft and Raumfahrt (1988/1989) had the same title; see Scheer, H. (ed) (1987) *Die gespeicherte Sonne: Wasserstoff als Lösung des Energie- and Umweltproblems*, Serie Piper, Munich; or more recently Rifkin, J. (2002) *Die H2-Revolution*, Campus Sachbuch.

8   Seifried, D. (1986) *Gute Argumente: Energie*, Munich.

9   Nitsch, J., Dienhart, H. and Langnis, O. (1997) 'Entwicklungsstrategien für solare energie-systeme. Die rolle von wasserstoff in Deutschland', in *Energiewirtschaftliche Tagesfragen*, vol 4, pp223–229; Dirschauer, W. (2003) 'Wasserstoff – kein königsweg der energieversorgung', *Energiewirtschaftliche Tagesfragen*, vols 1–2, pp87–89; Fischedick, M. and Nitsch, J. (2002) *Langfristszenario für eine Nachhaltige Energienutzung in Deutschland*, Forschungsbericht Bundes-ministerium für Umwelt, Berlin.

10 Pehnt,   M.   (2002)   *Energierevolution Brennstoffzelle? Perspektiven, Fakten, Anwendungen*, Weinheim 2002

11 For example, in the autumn of 2006, Taiwanese electronics firm Antig produced a charging device powered by a fuel cell for small appliances. The sales price of around US$2000, however, will mean that the device is only accepted in a small number of niche applica-tions; www.initiative-brennstoffzelle.de, 13 October 2006.

12 The utilities expect that they will own fuel cells, take the power on the grid, sell the heat, and thereby enter the heating market.

13 Vaillant's estimate, www.initiative-brennstof-fzelle.de (November 2006), IBZ-Nachrichten, December 2008; www.ibz-info.de

14 But things will not change overnight; there is also a lack of technical standards for hydro-gen lines to residential areas, for instance. *Deutscher Bundestag*, 14/5054.

15 *Deutscher Bundestag*, 14/5054, p102.

16 *Deutscher Bundestag*, 14/5054, p46.

17 Erste Brennstoffzellen-Flotte geht in den Praxistest, Dpa, 7 October 2002.

18 'Wasserstoffantrieb   als   Auslaufmodell', *Solarzeitalter*,   vol   4,   2009,   p106, www.eurosolar.org.

19 In   2006   the   German   Ministry   of Transportation estimated that the market maturity of fuel-cell cars will take about ten years, *Badische Zeitung*, 31 October 2006.

20 Langer, H. (2003) 'Brennstoffzellen-Tagung: Verhaltene   Äußerungen   zu   den Marktchancen', *Sonnenenergie*, January 2003, p. 34f. And there is another obstacle: switching vehicles to fuel cell drive trains will require considerable changes in industry (such as engine construction).

21 In   a   cost/benefit   analysis,   Germany's Environmental Protection Agency found that emissions reductions and resource protection would be much cheaper if classic drivetrains were optimized than if vehicles were equipped with fuel cells. *Deutscher Bundestag*, 14/5054, p65.

22 Measured as CO_2 equivalent.

23 The use of renewable electricity to produce hydrogen is counterproductive in terms of climate policy as long as the specific emissions of merit-order power are above 190g CO^2-eq/kWh. See Ramesohl, S., et al. (2003) *Bedeutung von Erdgas als neuer Kraftstoff im Kontext einer nachhaltigen Energieversorgung*, Wuppertal Institute for Climate, Environment and Energy, Wuppertal.

24 DaimlerChrysler (2003) *Umweltbericht*.

Chapter 9

1 DLR, IFEU and Wuppertal Institute (2004) *Ökologisch optimierter Ausbau der Nutzung erneuerbarer Energien in Deutschland. Forschungsvorhaben im Auftrag des Bundesministeriums für Umwelt, Naturschutz and Reaktorsicherheit*, Stuttgart, Heidelberg, Wuppertal. The authors make a distinction between two different types of potential. In the basic version, they determined the technical, structural potential that takes account of the main requirements for nature conservation, such as connections of wind farms in nature conservation areas. In the Nature Conservation Plus version, lower potential was determined under stricter nature conservation requirements (indicated in parentheses here). In the values indicated, a middle scenario was given for biomass, which can be used in a number of ways. In scenario A (100 per cent stationary use), the potential is 180TWh for electricity and 1130PJ of heat; in the other scenario (100 per cent fuels), all biomass use for fuels would amount up to 1000PJ.

2 In the version without the use of heat in cogeneration units, the figure is 290TWh; with heat usage, the potential of geothermal electricity drops to 66TWh.

3 BMU (2009) *Erneuerbare Energien in Zahlen*, Berlin, June 2009.

4 BMU (2006) *Erneuerbare Energien in Zahlen*, Berlin, May 2006.

5 www.solarwaerme-plus.info.

6 *Bundesverband Solarwirtschaft*, June 2006.

7 Resolution of the European Parliament on 13 March 2002; *Deutscher Bundestag* 14/9670.

8 Here, 16 systems were tested, 15 of which were found to be good or very good. *Stiftung Warentest*, vol 4, 2002.

9 Kommission der Europäischen Gemeinschaften, Energie für die Zukunft: Erneuerbare Energieträger, Weißbuch für eine Gemeinschaftsstrategie and Aktionsplan.

10 Kommission der Europäischen Gemeinschaften, Energie für die Zukunft: Erneuerbare Energieträger, Weißbuch für eine Gemeinschaftsstrategie and Aktionsplan, p8.

11 http://europa.eu/documents/comm/white_papers/pdf/com97_599_en.pdf.

12 Directive 2001/77/EC of the European Parliament and Council on 27 September 2001. Official Journal of the European Union L283/33 on 27 October 2001.

13 RL 2009/28/EG.

14 Data from 2002; Reiche, D. and Bechberger, M. (2005) 'Erneuerbare Energien in den EU-Staaten im Vergleich', *Energiewirtschaftliche Tagesfragen*, vol 10, pp732–739. The large share of hydropower also indicates why Sweden and Austria actually had lower shares of renewable power in 2003 then they had in 1997; less precipitation means less electricity from hydropower.

15 Reiche, D. and Bechberger, M. (2005) 'Erneuerbare Energien in den EU-Staaten im Vergleich', p732f; BMU (2006) *Erneuerbare Energien in Zahlen*, Berlin.

16 Reiche D. and Bechberger, M. (2005) 'Erneuerbare Energien in den EU-Staaten im Vergleich', p732f; BMU (2006) *Erneuerbare Energien in Zahlen*, Berlin.

17 BMU (2009) *Erneuerbare Energien in Zahlen*, Berlin, p51.

18 Reiche, D. and Bechberger, M. (2005) 'Erneuerbare Energien in den EU-Staaten im Vergleich', pp732–739.

19 Because Italy and Luxembourg passed new laws in March 2004, the European commission has not performed any reviews. Commission communiqué: The share of renewable energy in the EU. {SEK(2004) 547} 26 May 2004.

20 In a review of literature, researchers Ezzati, M. and Kammen, D. found that some 1.5–2 million deaths were the result of smoke from solid fuels worldwide in 2000. (2005) *REN21 Renewable Energy Policy Network: Renewable Energy 2005 Global Status Report*, Washington, DC, Worldwatch Institute, p29.

21 REN21 (2009) *Renewables Global Status Report*, 2009 Update.

22 China made up some 80 per cent of global growth in collector area in 2004. REN21 (2005) *Global Status Report 2005*, p25.

23 REN21 (2005) *Global Status Report 2005*, p25.

24 REN21 (2009) *Renewables Global Status Report*, 2009 Update.

25 Lehmann, H., et al. (1998) *Long-Term Integration of Renewables Energy Sources into the European Energy System*, The LTI Research Team, Physica Verlag.

26 Langniß, O., Luther, J., Nitsch, J. and Wiemken, E. (1997) *Strategien für eine nachhaltige Energieversorgung – Ein solares Langfristszenario für Deutschland*, Freiburg, Stuttgart.

27 DLR, IFEU and Wuppertal Institute (2004) *Ökologisch optimierter Ausbau der Nutzung erneuerbarer Energien in Deutschland. Forschungsvorhaben im Auftrag des Bundesministeriums für Umwelt, Naturschutz and Reaktorsicherheit*, Stuttgart, Heidelberg, Wuppertal.

28 In this calculation, power from renewables is compared to the primary energy that would have been needed to produce the same amount of electricity in thermal power plants (substitution method).

29 REN21 (2005) *Global Status Report*.

30 www.eee-info.net

31 DeNet (2009) *Wege in eine erneuerbare Zukunft*, Kassel, www.100-ee.de.

32 Köpke, R. (2004) 'Auf ewig heiß', *Neue Energie*, October, pp85–91.

33 ABC News (2008) 'Iceland moves to hydrogen power for ships', 23 January 2008, www.abc.net.au.

Chapter 10

1 Langniß, O. (1997) *Strategien für eine nachhaltige Energieversorgung – Ein solares Langfristszenario für Deutschland*, Freiburg/Stuttgart, p37.

2 Systems in which a steam turbine is hooked up behind the gas turbine. These systems are generally powered with gas or with gas and coal.

3 See glossary.

4 Lönker, O. (2005) 'Zukunftsspeicher', *Neue Energie*, April, pp30–36.

5 Engel, T. (2006) 'Der Ausbau erneuerbarer Energien braucht keine Speicher sondern Verbraucher', DGS newsletter, 2 November 2006.

6 www.thema-energie.de, the Alabama Electric Corporation has been operating a compressed air power plant since 1991 (110MW), in McIntosh, Alabama.

7 Lönker, O. (2005) 'Zukunftsspeicher', *Neue Energie*, April, p34.

8 Proposal made by Michael Vandenbergh (ISET), Engel, T. (2006) 'Der Ausbau erneuerbarer Energien braucht keine Speicher sondern Verbraucher', DGS newsletter, 2 November 2006. Another proposal is to use refrigerator systems to store electricity: 'At night, large refrigerator systems would reduce their temperature by a degree when energy consumption is low. But during the day, when energy consumption is high, they would increase their temperature by a degree. In the process, these cooling systems would function as batteries. Large refrigerator buildings have a storage capacity of 50,000MWh.' www.sonnenseite.com, 11 February 2007.

9 A fifth possibility is not considered because of the risks it entails: nuclear energy.

10 PROGNOS, DLR, Institute of Applied Ecology and Wuppertal Institute (1998) *Klimaschutzkonzept für das Saarland*, Untersuchung im Auftrag des Ministeriums für Umwelt, Energie and Verkehr des Saarlandes, Berlin, Stuttgart, Freiburg, Wuppertal.

11 Pick, E. and Wagner, H. J. (1998) 'Beitrag zum kumulierten Energieaufwand ausgewählter Windenergiekonverter', *Arbeitsbericht*, July 1998, Institut Ökologisch verträgliche Energiewirtschaft, Universität GH Essen

12 *Neue Energie*, September, p8 and www.dasgruene-emissionshaus.de.

13 Wagner, H. J. (2002) 'Energie- and Emissionsbilanz von Solaranlagen', *VDI Fortschrittberichte*, no 194, Düsseldorf, pp17–33; Stiftung Warentest, April

14 Kreutzmann, A. and Welter, P. (2005) 'Mehr raus als rein. Neue Energiebilanzen belegen kurze Energierücklaufzeiten von Photovoltaikanlagen', *Photon*, September 2005, pp66–69.

15 Here, the argument is that expanding renewables will increase retail electricity rates, thereby making the German industry slightly less competitive. The second aspect is lower demand. The money spent on a solar array cannot be used for other consumer items. As a result, jobs are lost in other sectors.

16 'Nachhaltige Energiewirtschaft, Einstieg in die Arbeitswelt von morgen', Institute of Applied Ecology, Freiburg, Darmstadt, Berlin 1996

17 *Wind Kraft Journal*, February, p44.

18 Because of the leading role that Denmark plays in this sector, Danish manufacturers have long dominated the global market for wind turbines (see 6.5).

19 BMU (2009) *Erneuerbare Energien in Zahlen*.

20 Kommission der Europäischen Gemeinschaften, Energie für die Zukunft: Erneuerbare Energieträger, Weißbuch für eine Gemeinschaftsstrategie and Aktionsplan, p15.

21 Seifried, D. (1986) *Gute Argumente: Energie*, Munich, p97.

22 Mez, L. (2001) 'Fusionen: Das große Fressen', *Energiedepesche*, June 2001

23 Verband der Deutschen Elektrizitätswirtschaft (VDEW): Strombilanz project group, 4 September 2006. See also Bömer, T. (2002) 'Nutzung erneuerbarer Energien zur Stromerzeugung im Jahr 2000', *Elektrizitätswirtschaft*, vol 7, pp22–32

24 BMU (2006) *Erneuerbare Energien in Zahlen*, Berlin.

Chapter 11

1 Over the years, various ministries have handled energy research, and the approaches have not always been clearly formulated. The categorization used here is the one set forth in the 3rd Energy Research and Energy Technologies Programme (1990–1995).

2 *Deutscher Bundestag*, 13/1963, p62ff, personal contact with Joachim Nitsch and with BMWi and BMBF.

3 BMBF (1996) 4rd Energy Research and Energy Technologies Programme, Bonn, p87.

4 *Deutscher Bundestag*, 14/8015.

5 Lönker, O. (2006) 'Forschen für Energie', *Neue Energie*, vol 4, pp22–27; BMWA (2005) 'Innovation and neue Energietechnologien', Das fünfte Energieforschungsprogramm der Bundesregierung, Berlin.

6 Lönker, O. (2006) 'Forschen für Energie', *Neue Energie*, vol 4, pp22–27.

7 Council decision of 18 December 2006 concerning the 7th Framework Programme of the European Atomic Energy Community (Euratom) for nuclear research and training activities (2007–2011) (2006/970/EURATOM).

8 Langniß, O., et al (1997) *Strategien für eine nachhaltige Energieversorgung – Ein solares Langfristszenario für Deutschland*, Freiburg/Stuttgart.

9 After the Cole compromise of 13 March 1997, the German federal government offered financial aid for the sale of electricity from coal and coal briquettes from 1997–2005 (nine years) in addition to funding for future shutdowns of coal mines in the amount of €28 billion. A further €1.3 billion was provided for the takeover of Saarbergwerke AG by Ruhrkohle AG. Overall, the subsidies amount to €3.3 billion per year during those nine years. In 2004, the governing coalition reached an agreement to extend the financing. From 2006–2012, the hard coal sector will receive just under €16 billion in public funding. In 2005, the sector received €2.8 billion; *Die Welt*, 9 November 2005.

10 According to German TV channel ARD, Germany consumes some eight million tons of kerosene per year. Each litre is exempted from roughly 60 cents of mineral tax in addition to sales tax.

11 A total of around €510 million over five years.

12 Umweltministerium and Wirtschaftsministerium Baden-Württemberg (2006) *Erneuerbare Energien in Baden-Württemberg 2005*, Stuttgart; preliminary values for feed-in rates in 2005: 45.4 billion kilowatt-hours, €4.4 billion.

13 Based on an average of 6 eurocents per kilowatt-hour.

14 Prognos AG (1992) *Die externen Kosten der Energieversorgung*, Stuttgart.

15 Hohmeyer, O. (1991) 'Least-Cost Planning and soziale Kosten', in Hennicke, P. (ed.) *Den Wettbewerb im Energiesektor planen*, Heidelberg, New-York, Tokyo. The external costs of conventional power generation are greater than the feed-in rates paid in Germany, as Hohmeyer shows in a report for Germany's Environmental Protection Agency. Hohmeyer, O. (2001) Vergleich externer Kosten zur Stromerzeugung in Bezug auf das Erneuerbare Energien Gesetz, Flensburg.

16 This is mainly due to the risk of malfunctions and the tremendous costs of a nuclear disaster. Up to now, damage has only been covered in insurance policies up to DM500 million – too little, as Ewers and Gelling reveal. The two researchers estimate that the possible damage amounts to around €2000 billion. And while such an amount can be ensured, the damage that such a disaster would have at an atomic plant could not be undone. Not even with trillions of euros can we bring people back to life or make radioactive land liveable again. In other words, even if we include the social costs of nuclear power generation, which Hohmeyer puts at 5–36 eurocents per kilowatt-hour, that only means that these costs can be attributed to atomic energy, not that the damage could be made undone with that money.

17 For instance, a study by DLR and the Fraunhöfer Institute for Systems and Innovation Research (ISI) found that the costs passed on to customers as part of the feed-in rates are lower than the external costs of power generated in fossil power plants.

DLR/ISI (2006) *Externe Kosten der Stromerzeugung aus erneuerbaren Energien im Vergleich zur Stromerzeugung*, Karlsruhe.

18 To the extent that the electricity tax does not exceed €511 per year.

19 At an overall efficiency of at least 70 per cent.

20 Ecologic, Institut für Internationale and Europäische Umweltpolitik gGmbH and Deutsches Institut für Wirtschaftsforschung (DIW) im Auftrag des Umweltbundesamtes: Auswirkungen der Ökologischen Steuerreform auf private Haushalte, Band III des Endberichts für das Vorhaben: 'Quantifizierung der Effekte der Ökologischen Steuerreform auf Umwelt, Beschäftigung and Innovation', August 2005.

21 Conze and Bitter (Stiebel Eltron) (1998) *Gesamtförderung 1997 von BUND*, Ländern, Kommunen und EVUs, 3 July 1998.

22 Lackschewitz, U. (GUT) (1998) *Thermische Solaranlagen in Wohngebäuden, Auswertung des Solarthermischen Förderprogramms des Landes Hessen für die Jahre 1992–1996*, Hessisches Ministerium für Umwelt, Energie, Jugend, Familie und Gesundheit, Kassel 1998, p7.

23 Lackschewitz, U. (GUT) (1998) *Thermische Solaranlagen in Wohngebäuden, Auswertung des Solarthermischen Förderprogramms des Landes Hessen für die Jahre 1992–1996*, p82.

24 Staiß, F. (2003) *Jahrbuch erneuerbarer Energien 2002/2003*, Bieberstein, p162ff; www.solarfoerderung.de

25 Lackschewitz, U. (GUT) (1998) *Thermische Solaranlagen in Wohngebäuden, Auswertung des Solarthermischen Förderprogramms des Landes Hessen für die Jahre 1992–1996*, p19; *Landtag Baden-Württemberg*, 12/1840, p5. The higher rate in the state of Hessen did not lead to greater growth; obviously, other factors are crucial for market development (such as general information and business activity).

26 *Landtag Baden-Württemberg*, 12/1840, 12/2340, 12/3635.

27 Lackschewitz, U. (GUT), (1999), ibid, p. 16 and p. 74; Landtag Baden-Württemberg 11/6318, II.1. and 12/242

28 Bleyl, J.; Krüger, K.: Solares Contracting. Sonnenenergie 1/2001, p. 24f. For further examples, see Sonnenenergie, 11/2002, p. 22f

29 As part of the German government's Solarthermie2000plus, several demonstration projects were conducted. Wärme für viele: Wohn- and Bürohäuser im Visier, Neue Energie, v. 2, 2007, p. 34

30 Erstes Gesetz zur Änderung des Gesetzes zur Förderung der sparsamen sowie umwelt- and sozialverträglichen Energieversorgung and Energienutzung im Land Berlin, 12 October 1995

31 „Leere Versprechungen?", Solarthemen no. 130, 28.2.2002, p. 2

32 "Berlin soll sich an Barcelona ein Beispiel nehmen" Neue Energie 2/2002, p. 52f; „Kein Neubau ohne Solaranlage", Sonnenenergie, March 2002, p. 48f.

33 For instance, the share of wind in total power consumption in Schleswig-Holstein exceeded 15 per cent in 1999, DEWI-Magazin, no 15, August 1999.

34 Gesetz für den Vorrang Erneuerbarer Energiequellen (Erneuerbare-Energien Gesetz), 21.7.2004, which went into effect in August 2004

35 Feed-in rates are passed on as a surcharge on top of retail electricity prices. Currently, the surcharge amounts to about 2 cents per kilowatt-hour. For large industrial power purchasers, whose electricity costs make up 15 per cent of gross added value or more, a special rule applies so that these consumers only pay an extra 0.05 cents per kilowatt-hour.

36 The feed-in rates applicable to the year of grid connection applies for 20 calendar years starting on the day of grid connection up to December 31 of the final year.

37 Compensation is subject to certain terms, see Section 6 EEG.

38 The floor rate for systems connected to the grid in subsequent years drops by 1 per cent per annum.

39 See Section 8 EEG, which specifies various compensation targets for electricity from biomass.

40 See Section 10 paragraph 3 EEG

41 *Global Status Report Renewables*, 2006 Update, p. 23

42 *Global Status Report Renewable Energy*, 2005, p. 21f

43 *Renewables Global Status Report* 2009

44 www.renewableenergyworld.com/rea/news/article/2009/07/britain-to-launch-innovative-feed-in-tariff-program-in-2010 and www.fitariffs.co.uk.

45 Manteuffel, B.: Size matters, Sonnenenergie, Nov. 2004, p. 16-19

46 For instance, the regional Schwarzwald- Baar-

Heuberg (Baden-Württemberg) Association warned of a blight on the landscape in August 2004. dpa, 18.8.2004

47 Section 11, paragraph 3 EEG

48 Photon, June 2006, p. 72

49 www.reuters.com/article/rbssUtilities Multiline/idUSN2050533620090320

50 BMU: Konsultationspapier zur Entwicklung eines Instruments zur Förderung der erneuerbaren Energien im Wärmemarkt. Berlin, 24 May 2006

51 Such contracts are private law agreements between the community and property purchasers in Germany. The section 23b of paragraph 9 of the Building Act serves as the basis; it specifies that "certain measures for the use of renewable energy" can be required.

52 'Kein Neubau ohne Solaranlage', Sonnenenergie, March 2002, p48f; for further information, see www.icaen.net.

53 Quilisch, T. and Peters, Chr. (2005) Solarthermie in Katalonien. Kommunale Solarverordnungen als Erfolgsmodell, Vortrag auf dem Exportforum, Hannover, 14 May 2005.

54 CDU-SPD coalition agreement in November 2005, p51.

55 BMU: Konsultationspapier zur Entwicklung eines Instruments zur Förderung der erneuerbaren Energien im Wärmemarkt. Berlin, 24 May 2006, www.bmu.de.

56 ibid.

57 EU directive 2003/87/EC.

58 Oberthür/Ott: The Kyoto-Protocol, Berlin, Heidelberg, New York 1999.

59 www.thema-energie.de and www.initiative-energieeffizienz.de.

60 RAVEL is an abbreviation for Rationelle Verwendung von Elektrizität. In this campaign, the Swiss focused on specific target groups for power conservation. Application-oriented research, further training for installers and engineers, demonstration projects, and awareness-raising were part of the campaign.

61 The reference showed how carbon emissions would have developed if nuclear plants have been left in operation.

62 Öko-Institut 1996: Das Energiewende-Szenario 2020. Ausstieg aus der Atomenergie, Einstieg in Klimaschutz and nachhaltige Entwicklung. Darmstadt, Freiburg, Berlin 1996.

63 International Solar Energy Society 2002:

Sustainable Energy Policy Concepts (SEPCo). Studie für das Bundesministerium für Umwelt, Naturschutz and Reaktorsicherheit, Freiburg 2002. For further information, see http://www.ises.org/shortcut.nsf/to/ICNFINAL.

Chapter 12

1 Forsa (2005) 'Meinungen zu erneuerbaren Energien', poll taken on 27/28 April 2005.

2 Forsa survey on renewables in January 2010, www.unendlich-viel-energie.de.

3 Güllner, M. (Forsa) (1999) 'Das Marktpotential and die Zielgruppe für "Öko-Strom"', Thesen im Rahmen des ASEW-Seminars: Top oder Flop? Ökostrom aus der Sicht der Stadtwerke, 27/28 April 1999 in Heidelberg.

4 www.ecotopten.de/prod_strom_prod.php.

5 From an ecological point of view, it does not matter whether clean power is generated simultaneously or earlier or later than it is consumed.

6 The Wuppertal Institute has come up with a concept for the assessment of an energy provider's sustainability services: 'Energieversorger auf dem Prüfstand', Wuppertal Papers no 116, 2001.

7 The figures given here represent the Freiburger Regio-Solarstromanlage: Wiese, R. (1997) 'Regio-Solarstromanlagen', Sonnenenergie, vol 5, pp34–35. For an overview of various business models, see Cottmann, T. (1997) 'PV-Betreibergemeinschaften: GmbH & Co KG?' Sonnenenergie, vol 5, p36f; Bernreuter, J. (2002) 'Mehr Risiko oder mehr Kosten', Photon, February 2002, pp36–39. The guide 'Bürger Solarstrom Anlagen', Verlag Solare Zukunft, Erlangen, is interesting for people who want to set up a community solar power array. See also www.bwebuecher. de.

8 For example, the Öko-Strombörse at the Bern Utility had 450kW of solar power under contract in the first year; Der Bund, 19 February 1999.

9 Elektrizitätswerke Zürich (EWZ), media conference on 15 April 1997.

10 It was called EKZ Naturstrom solar, www.ekz.ch/internet/ekz/de/kundendienst.

11 The Industrial Works of Basel offered such a programme, for example. But the programme was discontinued at the end of 2006.

12 www.eco-watt.de.

13 Seifried, D. (2008) Making Climate Change

Mitigation Pay: Eco-Watt: The Community-Financed Negawatt Power Plant, Freiburg.

14 '100,000W Solar-Initiative für Schulen in Nordrhein-Westfalen – Energieschule 2000+' is a project of Landesinitiative Zukunftsenergien NRW. The German state of North Rhine Westphalia promoted the project's design.

15 With hydraulics, the flow within a heating system is optimized. In principle, only the amount of water needed to heat a section of the building should flow through the pipes.

16 In this school, the heating and ventilation system was completely renovated, and a cogeneration unit and solar array, each with an electric output of 50kW, were installed; www.solarundspar.de.

17 www.solarundspar.de.

18 IEA 2006 (2006) *Light's Labour's Lost*, Paris, p25.

19 IEA 2006 (2006) *Light's Labour's Lost*, Paris, p468.

Glossary

Absorber
Part of a solar collector; see 2.1.

Air collectors
These devices use solar energy to heat air; see 2.5.

Auxiliary boiler
A boiler used alongside the principal boiler (such as a wood-fired boiler) to provide extra heat on days of peak demand.

Baseload
The level of power that is needed all the time, day and night, summer and winter; also see 'peak load'.

Biogas
Gas created in an oxygen-free environment when biomass (especially excrement) ferments; see 5.2 and 5.3.

Biomass
See Chapter 5.

Climate policy
Policies designed to slow down or prevent global warming (see 'greenhouse effect'). Because carbon emissions are a main cause of the greenhouse effect, climate policies often aim to reduce carbon emissions.

Cogeneration unit
A system that generates electricity but also supplies its waste heat as useful heat. For instance, a (gas) motor can be used to drive a generator. In addition to the electricity, the waste heat from the motor would also be used; see 1.9 and 5.3.

Collectors
Convert sunlight into heat; see 2.1.

Community projects
See 12.3.

Condensation boiler
A gas or oil heater with especially great efficiency. The water vapour in the waste gas condenses, thereby increasing overall efficiency.

Contracting
A way of financing a new, more efficient system (such as lighting or ventilation) with future savings from lower energy costs. The contractor plans the investment, then finances and implements it. In return, the contractor receives the difference between the old and the new energy costs over a specified period; see 12.6 and 12.7.

Conventional power
Power from a central plant whose waste heat from power production simply escapes into the atmosphere; such plants generally have a relatively low efficiency of around 30 to 45 per cent.

Demand management
When demand for power is greater than the capacity of power plants, there are two solutions: either power plant capacity can be increased, or some power consumers can be briefly switched off. The second option is called demand management.

Diffuse/direct sunlight
The sunlight that reaches the Earth in a cloudless sky from a particular angle is called direct sunlight. In contrast, diffuse sunlight is refracted by clouds, smog, fog, etc. and it reaches the Earth's surface from different angles; see 1.6.

District heating network

An insulated network of pipes which supply heat from a central heating/solar system to consumers nearby; see 2.3 and 5.5.

Eco-taxation

A tax on the consumption of resources; in Germany, the tax has been applied to oil, natural gas, petrol and electricity; see 11.4 and 11.5.

Efficiency

The ratio between energy input and energy output. For instance, if 100,000kWh of energy is put into a power plant as fuel (energy input), and the power plant uses that energy to generate 35,000kWh of electricity, the efficiency is 0.35 or 35 per cent.

Electrolysis

When electricity is used to break up a chemical compound; for instance, electrolysis can split water into oxygen and hydrogen.

Energy-conservation ordinance

This German law currently sets a limit on energy consumption in buildings; see 4.1.

Energy crops

Plants grown especially as energy sources; see 5.1 and 5.6.

Feed-in Act

A German law obligating grid operators to pay a floor price for renewable electricity exported to their grids. In 2000, it was replaced by the Renewable Energy Act (EEG); see 11.9.

Feed-in rates

A floor price paid for electricity exported to the grid.

Final energy

The energy provided to consumers (homes, industry, vehicles, etc.); in other words, the energy provided after primary energy has been converted in refineries, power plants, etc. and distributed over the electricity grid, filling stations, etc.

Forestry waste

Wood collected in forest management that is too small to be used in sawmills. It can, however, still be used to make particle board or as a source of energy (wood chips and wood pellets); see Chapter 5.

Fossil energy sources

Coal, petroleum and natural gas, which contain solar energy stored from previous millennia; they are all the products of geologically captured plant matter.

Fuel cell

A source of electricity (and heat) in which a fuel's chemical energy (such as hydrogen or natural gas) is directly converted into electricity. The heat created in the process can also be used.

Gas turbine

A special turbine that can easily be ramped up and down; see 10.1.

Geothermal

The use of underground heat as a source of energy.

Green power

Electricity from renewable energy; see 12.2.

Greenhouse effect

Heat-trapping gases in the Earth's atmosphere (mainly carbon dioxide, methane, laughing gas, etc.) prevent heat from escaping into outer space. As a result, the Earth's atmosphere heats up; see 1.1.

Grid-connected solar arrays

Photovoltaic systems which do not have a battery to store solar energy. The power that is not consumed locally is exported to the grid; see 3.2.

Grid transit fees

The fees that are charged when electricity travels through another company's grid.

Heat Conservation Ordinance

A German law that specified maximum heating demand in new buildings; in 2002 it was replaced by the Energy Conservation Ordinance; see 4.1.

Heat recovery

In buildings, heat exchangers are often integrated into ventilation systems. They take energy out of the warm outgoing air and use it to preheat colder incoming air. The efficiency of modern heat exchangers in such systems exceeds 90 per cent.

Insulating glass

Window panes that are good insulators; see 4.2.

Liberalization of the energy market

See 1.10.

Low-energy house

A house with low heating energy demand; see 4.1.

Municipal utilities

Utility companies that provide electricity, gas and possibly water generally within the city limits

Negawatts

See 12.7.

Nuclear fusion

In nuclear fusion, energy is generated when nuclear atoms melt. At present, nuclear fusion is a long way from being a viable technical application (see 11.1).

Off-grid solar house

See 4.7.

Offshore turbines

Wind turbines installed in shallow water offshore.

Passive house

A house with very low heating energy demand; it can even do without a conventional heating system; see 4.6.

Passive solar energy

See 4.2.

Payback

The time it takes for an investment to pay for itself. Energy payback is the time a renewable energy system needs to generate the energy used for its manufacture; see 10.4.

Peak load

Times when demand for power is at its highest; generally, the demand only lasts a few hours or a few days a year; see 'baseload'.

Photovoltaics

These systems directly convert solar energy into electricity; see Chapter 3.

Plus-energy house

A house that generates more energy than it consumes; see 4.8.

Primary energy

Coal, natural gas, petroleum, wind, flowing water, sunlight, biomass and geothermal heat are primary sources of energy. The energy contained in them is called primary energy. In contrast, electricity is a secondary type of energy because it does not directly occur in nature, but must first be created from primary energy sources.

Pumped storage

See 'storage power plant'.

Quota model

See 11.13.

Renewable Energy Act (EEG)
The German law that replaced the Feed-in Act; see 11.10.

Run-of-river power plant
Hydro plants that use flowing river water to generate electricity. Generally, there is little difference between the level of water on either side of the dam, but the amount of water is great. For economic reasons, such dams are generally constructed with sluices.

Scenario
A model for the future based on certain assumptions and mathematical equations; see 9.6.

Service water
Hot water used in bathrooms and kitchens.

Solar hydrogen
Hydrogen made from solar energy as a way of storing energy; see 8.2.

Solar ordinance
An ordinance that obligates some or all builders to install solar arrays; see 11.14.

Solar power exchange
An agency that brokers between buyers and sellers; see 12.4.

Solar thermal
The use of solar energy to create heat; see Chapter 2.

Solar thermal power plants
Power plants that use solar energy to heat up on the heat carrier (such as water or another fluid) to generate electricity with a normal steam-driven generator; see 2.6.

Storage power plant
Hydro plants that utilize differences in altitude, such as around mountain lakes. Pumped storage plants are a special type; here, inexpensive electricity (such as power at night) is used to pump water up to a storage basin, and when power is needed during peak demand, the water can be let through to drive a turbine and generate electricity at times when power is more expensive; see 10.2.

Sustainable
An economic activity is considered sustainable if it does not consume resources faster than they can regenerate.

Transmission losses
Heat losses that occur when energy passes through parts of the house, such as walls, the roof and windows. Insulation reduces transmission losses.

Transparent insulation
Special insulation applied to the outside of a house to capture solar energy; see 4.5.

Wind farm
A project consisting of numerous wind turbines.

Woodchip heating
Heating systems fired with timber chopped up into chips; see 5.5.

Wood pellets
A special fuel for wood heating systems. Wood pellets are some 6–12mm long and generally consist of waste wood products; the pellets themselves have standardized properties, such as moisture content and heating value; see 5.4.

Index